U.S. Engineering in a Global Economy

**A National Bureau
of Economic Research
Conference Report**

U.S. Engineering in a Global Economy

Edited by **Richard B. Freeman and Hal Salzman**

The University of Chicago Press

Chicago and London

The University of Chicago Press, Chicago 60637
The University of Chicago Press, Ltd., London
© 2018 by the National Bureau of Economic Research
Published 2018
Printed in the United States of America

27 26 25 24 23 22 21 20 19 18   1 2 3 4 5

ISBN-13: 978-0-226-46833-4 (cloth)
ISBN-13: 978-0-226-46847-1 (e-book)
DOI: https://doi.org/10.7208/chicago/9780226468471.001.0001

Library of Congress Cataloging-in-Publication Data

Names: Freeman, Richard B. (Richard Barry), 1943– editor. | Salzman,
    Hal, editor.
Title: U.S. engineering in a global economy / edited by Richard B.
    Freeman and Hal Salzman.
Other titles: National Bureau of Economic Research conference report.
Description: Chicago ; London : The University of Chicago Press,
    2018. | Series: National Bureau of Economic Research conference
    report
Identifiers: LCCN 2017040609 | ISBN 9780226468334 (cloth : alk.
    paper) | ISBN 9780226468471 (e-book)
Subjects: LCSH: Engineers—Employment—United States.
Classification: LCC TA157 .U845 2018 | DDC 331.7/62000973—dc23
LC record available at https://lccn.loc.gov/2017040609

♾ This paper meets the requirements of ANSI/NISO Z39.48-1992
(Permanence of Paper).

Relation of the Directors to the
Work and Publications of the
National Bureau of Economic Research

1. The object of the NBER is to ascertain and present to the economics profession, and to the public more generally, important economic facts and their interpretation in a scientific manner without policy recommendations. The Board of Directors is charged with the responsibility of ensuring that the work of the NBER is carried on in strict conformity with this object.

2. The President shall establish an internal review process to ensure that book manuscripts proposed for publication DO NOT contain policy recommendations. This shall apply both to the proceedings of conferences and to manuscripts by a single author or by one or more co-authors but shall not apply to authors of comments at NBER conferences who are not NBER affiliates.

3. No book manuscript reporting research shall be published by the NBER until the President has sent to each member of the Board a notice that a manuscript is recommended for publication and that in the President's opinion it is suitable for publication in accordance with the above principles of the NBER. Such notification will include a table of contents and an abstract or summary of the manuscript's content, a list of contributors if applicable, and a response form for use by Directors who desire a copy of the manuscript for review. Each manuscript shall contain a summary drawing attention to the nature and treatment of the problem studied and the main conclusions reached.

4. No volume shall be published until forty-five days have elapsed from the above notification of intention to publish it. During this period a copy shall be sent to any Director requesting it, and if any Director objects to publication on the grounds that the manuscript contains policy recommendations, the objection will be presented to the author(s) or editor(s). In case of dispute, all members of the Board shall be notified, and the President shall appoint an ad hoc committee of the Board to decide the matter; thirty days additional shall be granted for this purpose.

5. The President shall present annually to the Board a report describing the internal manuscript review process, any objections made by Directors before publication or by anyone after publication, any disputes about such matters, and how they were handled.

6. Publications of the NBER issued for informational purposes concerning the work of the Bureau, or issued to inform the public of the activities at the Bureau, including but not limited to the NBER Digest and Reporter, shall be consistent with the object stated in paragraph 1. They shall contain a specific disclaimer noting that they have not passed through the review procedures required in this resolution. The Executive Committee of the Board is charged with the review of all such publications from time to time.

7. NBER working papers and manuscripts distributed on the Bureau's web site are not deemed to be publications for the purpose of this resolution, but they shall be consistent with the object stated in paragraph 1. Working papers shall contain a specific disclaimer noting that they have not passed through the review procedures required in this resolution. The NBER's web site shall contain a similar disclaimer. The President shall establish an internal review process to ensure that the working papers and the web site do not contain policy recommendations, and shall report annually to the Board on this process and any concerns raised in connection with it.

8. Unless otherwise determined by the Board or exempted by the terms of paragraphs 6 and 7, a copy of this resolution shall be printed in each NBER publication as described in paragraph 2 above.

This volume is dedicated to Ralph Gomory for his encouragement of scientific analysis of the job market for scientists and engineers and his contribution to economic analysis of technology, international trade, and firm and industry competitiveness.

Contents

Acknowledgment

We gratefully acknowledge funding from the Alfred P. Sloan Foundation.

Introduction

Richard B. Freeman and Hal Salzman

From the late 1950s—when the launch of Sputnik produced fears that the United States was losing its technological leadership to the Union of Soviet Socialist Republics (USSR)—to the present, the state of the labor market for specialists in STEM (science, technology, engineering, and mathematics) occupations has attracted considerable public attention, spurring analysis in workforce development, labor economics, and economics more broadly. Public concern historically has focused on possible shortages of scientists and engineers hampering economic growth or national security. The National Bureau of Economic Research (NBER) has a long history of analyzing the science and engineering workforce, beginning with Blank and Stigler's (1957) *The Demand and Supply of Scientific Personnel*, which came in the same year of the Sputnik launch. While the title referred to scientific personnel, most of the book dealt with the engineering profession. This is not surprising since the vast majority of STEM workers in industry were, at that time, engineers. Today, the expansion of the biomedical workforce and of computer science and other information technology (IT) workers has overtaken the numeric dominance of engineering among STEM workers, but engineering remains a critical part of scientific personnel in industry and the largest number of STEM workers in many industries.

The central issue in the labor market analysis of engineers following Sput-

Richard B. Freeman holds the Herbert Ascherman Chair in Economics at Harvard University and is a research associate and director of the Sloan Science Engineering Workforce Projects at the National Bureau of Economic Research. Hal Salzman is Professor of Planning and Public Policy at the Edward J. Bloustein School of Planning & Public Policy and Senior Faculty Fellow at the John J. Heldrich Center for Workforce Development at Rutgers University.

For acknowledgments, sources of research support, and disclosure of the authors' material financial relationships, if any, please see http://www.nber.org/chapters/c12683.ack.

nik was the meaning of a shortage in a flexible market economy where wages and prices move freely to clear supply and demand. When demand for labor rises relative to supply, wages rise and the supply increases, so what exactly is a shortage? Arrow and Capron (1959) treated shortages as the result of rapid shifts in demand, such as the huge increase in demand for engineers, physicists, and others sparked by the U.S. effort to surpass the USSR in space technology. Freeman (1971) put the supply and demand responses together into a cobweb model in which cyclical fluctuations in wages, employment, and enrollments arise naturally from the lag in supply responses, due to the years students spend learning STEM skills. Engineering was the prime exemplar of this pattern.

In the 1980s and 1990s, NBER work by Zvi Griliches (1998) and Griliches and Lichtenberg (1984) examined the link between research and development (R&D) spending and private-sector productivity in a production-function framework. While the econometrics of adding R&D spending to production functions may seem far removed from the labor market, the analysis can be viewed as an investigation of the demand side of the science and engineering market. About three-quarters of R&D spending consists of wages and salaries of scientists and engineers, and the derivative of the production function with respect to the number of scientists and engineers is the derived demand for those workers.

In the late 1990s and early in the twenty-first century, NBER studies of science and engineering examined the inconsistency between wage, employment, and enrollments data and what seemed like perpetual claims of shortages. Eric Weinstein (2003) analyzed the misuse of evidence on supply and demand data behind some of the 1980s alarmist cries of shortages from major company leaders and government officials, including the National Science Foundation, for which the agency's head eventually apologized. Teitelbaum (2014) documents the history of shortage claims through the first decade of the twenty-first century. Looking outside the shortage debate, Austen Goolsbee (1998) asked whether government R&D policy largely benefited scientists and engineers by driving up their salaries, while Paul Romer (2000) examined the benefits and costs of government subsidies of R&D.

From the first decade of the twenty-first century to the present, research continued on the productivity effects of R&D (see Hall, Mairesse, and Mohnen [2009], among others), but a different set of issues came to the fore in labor market analysis. On the demand side, Lynn Zucker and Michael Darby (2006) looked at the effects of the location of top scientists and engineers on the formation of high-tech firms. On the supply side, Richard Freeman (2005) examined the globalization of the science and engineering workforce and its potential effects on the future position of the United States in the global economy. With Sloan Foundation support, the NBER set up the Science and Engineering Workforce Project that primarily focused on the doctoral workforce in the academic sector, as reported in Freeman and Goroff (2009). Recognizing the increased importance of immigrants and women in the STEM workforce, Jenny Hunt examined where immigrant engineers fit in the education and earnings distribution of engineers (Hunt 2010) and the factors that lead women to leave engineering and science more quickly than men (Hunt 2013).

U.S. Engineering in a Global Economy follows the NBER tradition of quantitative analysis of the demand and supply sides of the engineering job market in the United States. Many of the chapters use novel data or approaches to examine engineering education, practice, and careers in ways designed to inform science and engineering educational institutions, funding agencies, and policymakers about the challenges of developing and employing engineers in ways that most efficaciously contribute to the innovation driving modern economic growth.

Chapter 1 sets the stage for the rest of the book with a review of the engineering labor force, focusing on the employment, salary, and career trajectories of graduates that obtain engineering degrees and work in the field. Lacking a single comprehensive data source on engineers, this chapter draws on a wide variety of longitudinal career data and establishment-based employment and earnings data. These come from different government surveys of scientists, engineers, and employers, including census survey data on the numbers from overseas, and from education administrative data on the supply of engineers coming from U.S. universities. It disaggregates engineering into major subfields, whose employment differs sufficiently to face different supply and demand conditions.

Chapters 2, 3, and 4 focus on supply issues. In chapter 2, Gilmartin, Antonio, Brunhaver, Chen, and Sheppard use a fifty-item survey instrument administered to over 4,000 students across twenty-one U.S. colleges and universities to examine the educational pathways through which junior and senior engineering students move from school to the labor market. They examine the correlations of their postgraduation plans, their psychological motivations, and the attributes of the programs in which they may major. In chapter 3, Weinberger merges data on degrees in historically black colleges and universities (HBCUs) with labor force data to analyze the geography and timing of the increased supply of minority graduates into the STEM fields, giving special attention to how the historically black institutions responded to increased opportunities for blacks in engineering when business reduced discriminatory barriers. Chapter 4 presents Brunhaver, Korte, Barley, and Sheppard's analysis of the experience of engineering students who transitioned from their studies to engineering workplaces. Examining the skills the graduates used at work and where they learned those skills, they provide insight into the strengths and weaknesses of educational programs and on-the-job training that economists usually measure simply as years of work experience.

Chapters 5 and 6 turn to the demand side of the market, using different forms of data to investigate the contribution of engineers to productivity and innovation. In chapter 5 Barth, Davis, Freeman, and Wang combine establishment-level production data with firm-level R&D data and census data on the occupations of workers to estimate the contribution of scientists and engineers working outside of R&D labs on the productivity at their workplace. In chapter 6, Helper and Kuan report the results of a survey of over one thousand firms in the automobile supply chain and results of interviews of dozens of engineers, workers, and managers on the contribution of incremental innovations of small suppliers to the growth of productivity that national statistics measure only in final product data. The two chapters mesh together well, as the Helper and Kuan interviews and surveys provide valuable insight into interpreting the statistical calculations of Barth and colleagues in chapter 5.

The last three chapters deal with the operation of engineering labor markets. The United States and most other advanced countries use some form of occupational licensing to ensure that persons practicing in the field have requisite training and skills. Hur, Kleiner, and Wang give a detailed empirical analysis of occupational licensing in civil, electrical, and industrial engineering and its impacts on earnings and employment in chapter 7. In the tradition of the Freeman cobweb model of the interaction of supply and demand, Lynn, Salzman, and Kuehn show in chapter 8 the response of universities and students to an upswing in demand for petroleum engineers that highlights the large elasticity of the domestic labor supply to sharp increases in wages. Examining the increased use of foreign overseas supply of engineers in the United States, Hira uses data from the U.S. Departments of Labor and Homeland Security in chapter 9 to analyze the differences between firms that use the H-1B program to provide lower-cost temporary labor and those using the program as a bridge toward getting permanent immigration status for employees.

Each of the chapters gives a detailed report of the data used, the methodology applied, and the findings. The range of data used to illuminate the job market is wide, from special surveys of graduate students, programs, and firms to administrative data, government surveys, industry, and engineering association reports, licensing and visas, to news reports of firm attitudes and concerns about visas. There is a smorgasbord of information in the chapters and a wide range of references to work in different areas and from different disciplines. To see the linkages among the different studies and the ways in which findings fit together, we summarize below what we view as the three overarching themes from the book.

Supply, Demand, and Globalization

The supply of engineers to the U.S. labor market is responsive to economic conditions because students and engineering programs pay attention to economic signals and globalization provides new channels of supply.

Three of the chapters give evidence of the supply responsiveness by students and universities that gainsay the view that the U.S. supply system functions poorly and that underlies the perennial warnings about shortages of scientists and engineers. The findings of these chapters provide strong evidence that students and the educational institutions that prepare them for careers in science and engineering are aware of economic opportunities and are quick to respond to these opportunities and market signals.

The strongest evidence of sizable supply responses are given by analyses of the flow of students and by changes in university programs responding to market conditions. Given historically limited opportunities for black graduates in the private sector, relatively few blacks became engineers and historically black colleges provided limited educational offerings in engineering. Weinberger's analysis shows that when the barriers of discrimination lowered, businesses, foundations, and HBCUs made a concerted effort to expand educational opportunities in engineering, computer science, and other technical fields, "to prepare their students for expanded career choices." Students responded and the result was a substantial increase in the number of college-educated black men and women entering engineering, particularly from the six HBCUs that were in the forefront working with businesses and foundations. Treating the opening of new programs as a supply-side shock to educational opportunity, Weinberger finds that the graduates who went into these STEM fields had better labor market outcomes than those in other occupations or in earlier birth cohorts.

The Lynn, Salzman, and Kuehn study documents the responses to "a quasi-natural experiment" in petroleum engineering when, early in the twenty-first century, demand for that specialty increased greatly after decades of little hiring. Industry raised entry-level wages, and within two to three years the number of graduates in petroleum engineering began increasing so rapidly that by 2015 the number of graduates was five times the number in 2005–2006! Even in a very specialized field, supply is highly responsive to traditional market signals of wages. Interviews with department chairs and others show the effort by academic institutions to increase supply so as to meet the market demand.

The Gilmartin, Antonio, Brunhaver, Chen, and Sheppard analysis of students who major in engineering gives a more nuanced picture of supply behavior. It finds that "over two-thirds [of engineering students] having non-engineering, mixed, or uncertain plans," and these students differ from the students with engineering-focused career plans based on modest differences in median salaries in their region. It is notable that engineers show greater flexibility for their future work plans than one might have expected from such specialized education, "with over two-thirds having nonengineering, mixed, or uncertain plans." The openness that students show to pursue pathways outside of engineering is consistent with evidence that about one-third of the 70,000–75,000 engineering graduates in the United States each year take nonengineering jobs because they report finding other careers more attractive (Salzman, Kuehn, and Lowell 2013).

Productivity and Innovation

Engineers and scientists outside of formal R&D raise productivity both in their company and through innovations along the supply chain to places beyond their employer.

Three chapters use different types of data to give evidence on the link between what scientists and engineers do outside of formal R&D activities and productivity. It is important to analyze what these non-R&D scientists and engineers do because they make up the majority of persons in science and engineering occupations. Between 70 and 80 percent of scientists and engineers in U.S. industry work on non-R&D activities. At the doctoral level, 45 percent of all PhDs in the industry report that their work does not include research as a primary or secondary activity. Traditional production-function analyses that make R&D the key determinant of labor or total factor productivity neglect the possible contribution these scientists and engineers make to output by implementing or improving new technologies and the impact of such improvements along the supply chain to other firms.

Barth, Davis, Freeman, and Wang's production-function investigation shows that in manufacturing, establishments that have higher proportions of scientists and engineers have higher productivity in both cross-section comparisons of establishments and, perhaps more convincingly, in comparisons of the same establishment when it changes the proportion of its workforce in science and engineering over time. The evidence further suggests that the effects of having more scientists and engineers at establishments is larger, the greater the intensity of R&D activity. Some of the benefits from higher productivity appear, moreover, to spill over to higher earnings for non-STEM workers.

The Helper and Kuan surveys and interviews show that many non-R&D engineers contribute to the introduction of new products or processes and/or to lowering costs of production, providing examples of both effects. They find that engineers at supplier firms in the automobile sector contribute many incremental gains that would not meet the term "innovation" nor fit under any R&D rubric, and that many work closely with customers in generating improvements. Further, some non-R&D engineers work closely with production workers and thus jointly contribute to productivity improvements. Their findings support the notion that standard production functions that focus solely on the R&D pathway to technological progress do not capture the reality of how non-R&D engineers contribute to productivity. This analysis also shows wide heterogeneity in firm policies and practices, even within the same detailed industry classification.

Finally, Brunhaver, Korte, Barley, and Sheppard's interviews with early career engineers about their actual work gives further insight into what engineers do outside formal R&D facilities. These engineers report their work as more variable *and* complex than academic curricula convey. Moreover, their work often "is less about using theories or equations, for example, than about project management and working with other people." While these interviews did not probe into how (if at all) their work raised productivity, it shows the importance of nontechnical skills even for beginning engineers in industry. Parenthetically, it also fits with the openness that engineering students have toward alternative career paths and curriculum reforms that broaden the scope of skills that make up an engineering degree, and with other studies of employers saying the nonengineering skills are those that are the harder to find and the more sought after skills of new graduates (Lynn and Salzman 2010).

The findings in these three chapters from different data, enterprises, and industries provide breadth in examining aspirations and plans of young engineers, the productivity outcomes from the work of incumbent scientists and engineers, and how engineers outside of R&D improve efficiency at their own firm and across different points in the supply chain. They show consistency in the conclusion that engineers contribute to productivity and innovation much more broadly than recognized in formal models of R&D activity.

Education and Labor Markets

There is considerable variation in the way the institutional structure of education and labor markets affects outcomes.

Market outcomes depend not only on the classic forces of supply and demand but also on the institutional or legal structure that influence the decisions or that determine outcomes through law or regulation. Two chapters of the book examine how laws and regulations affect the engineering job market.

Hur, Kleiner, and Wang's analysis of the licensing of engineers shows that it has expanded over time with, however, large variation in its existence and strictness across states that, surprisingly, is unrelated to the usual state policies regarding labor regulations. States with the most restrictive licenses included Georgia, Texas, Pennsylvania, and Illinois, while the states with the least restrictive licensing laws were Virginia and Minnesota. But regulations had small and often insignificant impacts on wages or hours worked, implying that market forces dominated the nature of licensing. While not the main focus of Lynn, Salzman, and Kuehn's analysis of petroleum engineering, they also found that market forces in higher education overwhelmed the efforts by two leading petroleum engineering departments, Texas A&M and University of Texas at Austin, to moderate student supply responses in the hope of avoiding an excessive increase in supply when wages increased for petroleum engineers. However, the dramatic increase in wages induced enough other departments to admit students to create the fivefold increase noted earlier, resulting in graduating more engineers than industry was hiring, and coincided with a decline in oil prices that further depressed demand. Taken together, these two studies show that broad market forces are sufficiently strong to overwhelm the effects of states acting individually through licensure and of large departments acting individually in their admission policies to have any noticeable effect on outcomes.

Hira's analysis of H-1B visas tells a more complex story about the interaction between institutions and market forces. On the one side, the H-1B law determines the number of visas for temporary migrants and thus bounds the supply of such workers. By lodging control of the visas with the employer, the H-1B program further segregates H-1B migrants from the general labor market for engineers. The H-1B recipients cannot change employers and use the normal channel of job changes or threats of changes to improve their economic position, which assures that employers are major beneficiaries of the program. On the other side, the market forces and company strategies in different parts of the IT industry lead firms to use the visas in different ways within the same institutional framework. One set of employers uses the H-1B program solely for getting work done at low wages, hiring foreign workers with no effort to sponsor them for permanent residency. Another set of employers pays higher wages to their H-1B workers and appears to use the H-1B program as a way of selecting some workers to integrate into their permanent workforce by sponsoring them for permanent residency.

In sum, the book offers insight into a variety of issues in the changing market for engineers and highlights others that might be fruitfully addressed in future research. We need to know more about the actual work activity of persons with engineering and other STEM degrees not only outside R&D, which the book deals with, but outside science or engineering entirely to get a full picture of the value of this formal education, and of ways to improve the link from schooling to work. We also need to better understand the ways in which firms, students, and training institutions respond to a global market in which U.S. workers and firms face competition unlike that which we have had in the past. In addition, we need insight into the best ways to improve science and engineering education to fit current and future demands of the workforce and, as always in economics, about the wide heterogeneity of labor market outcomes among workers and firms and their relation to explicit policies and regulations.

References

Arrow, Kenneth, and William Capron. 1959. "Dynamic Shortages and Price Rises: The Engineer-Scientist Case." *Quarterly Journal of Economics* 73 (2): 292–308.

Blank, David, and George Stigler. 1957. *The Demand and Supply of Scientific Personnel*. Cambridge, MA: National Bureau of Economic Research.

Freeman, Richard. 1971. *The Market for College-Trained Manpower*. Cambridge, MA: Harvard University Press.

———. 2005. "Does Globalization of the Scientific/Engineering Workforce Threaten U.S. Economic Leadership?" NBER Working Paper no. 11457, Cambridge, MA.

Freeman, Richard, and Daniel Goroff, eds. 2009. *Science and Engineering Careers in the United States: An Analysis of Markets and Employment*. Chicago: University of Chicago Press.

Goolsbee, Austen. 1998. "Does Government R&D Policy Mainly Benefit Scientists and Engineers?" NBER Working Paper no. 6532, Cambridge, MA.

Griliches, Zvi. 1998. *R&D and Productivity: The Econometric Evidence*. Chicago: University of Chicago Press.

Griliches, Zvi, and Frank Lichtenberg. 1984. "Interindustry Technology Flows and Productivity: A Reexamination." *Review of Economics and Statistics* 22 (2): 324–29.

Hall, Bronwyn, Jacques Mairesse, and Pierre Mohnen. 2009. "Measuring the Returns to R&D." NBER Working Paper no. 15622, Cambridge, MA.

Hunt, Jennifer. 2010. "Why Do Women Leave Science and Engineering?" NBER Working Paper no. 15853, Cambridge, MA.

———. 2013. "Are Immigrants the Best and Brightest U.S. Engineers?" NBER Working Paper no. 18696, Cambridge, MA.

Lynn, Leonard, and Hal Salzman. 2010. "The Globalization of Technology Development: Implications for U.S. Skills Policy." In *A U.S. Skills System for the 21st Century: Innovations in Workforce Education and Development*, edited by D. Finegold, M. Gatta, H. Salzman, and S. Shurman. Champaign: Labor and Employment Relations Association, University of Illinois at Urbana-Champaign.

Romer, Paul. 2000. "Should the Government Subsidize Supply or Demand in the Market for Scientists and Engineers?" NBER Working Paper no. 7723, Cambridge, MA.

Salzman, Hal, Daniel Kuehn, and B. Lindsay Lowell. 2013. "Guestworkers in the High-Skill U.S. Labor Market: An Analysis of Supply, Employment, and Wage Trends." Briefing Paper no. 359, Washington, D.C., Economic Policy Institute.

Teitelbaum, Michael. 2014. *Falling Behind? Boom, Bust, and the Global Race for Scientific Talent*. Princeton, NJ: Princeton University Press.

Weinstein, Eric. 2003. "How and Why Government, Universities, and Industry Create Domestic Labor Shortages of Scientists and High-Tech Workers." NBER Working Draft, Cambridge, MA. https://www.ineteconomics.org/uploads/papers/Weinstein-GUI_NSF_SG_Complete_INET.pdf.

Zucker, Lynn, and Michael Darby. 2006. "Movement of Star Scientists and Engineers and High-Tech Firm Entry." NBER Working Paper no. 12172, Cambridge, MA.

1

The Engineering Labor Market
An Overview of Recent Trends

Daniel Kuehn and Hal Salzman

1.1 Introduction

The role of engineers in developing infrastructure, technology, and innovation has long made the economic health of the profession an issue of concern for public policy. In the Great Depression, engineers were critical in developing the New Deal public infrastructure that boosted employment. In World War II, engineers were critical in advancing military technology and deploying it in the field, which raised national security concerns about bottlenecks due to limited supply (Allen and Thomas 1939). After the war, many saw engineers and scientists as vital to national security and economic prosperity. In 1945, Vannevar Bush (an engineer himself) articulated the need for a strong science and engineering workforce in his famous statement on the "Endless Frontier" of scientific and technological progress (Bush 1945). For most of the next several decades, policymakers worried about shortages of engineers and scientists.[1] Shortage fears reached a crescendo in 1957, when the Soviet Union's launch of Sputnik seemed to threaten U.S.

Daniel Kuehn is a research associate at The Urban Institute. Hal Salzman is Professor of Planning and Public Policy at the Edward J. Bloustein School of Planning & Public Policy and Senior Faculty Fellow at the John J. Heldrich Center for Workforce Development at Rutgers University.

We are grateful for insights provided by Catherine Didion, Richard Freeman, Ron Hira, Robert Lerman, Mark Regets, Proctor Reid, and Sheri Sheppard, participants at the National Bureau of Economic Research/Sloan Engineering Project Workshop, and Urban Institute seminar participants. This chapter is based on research supported by the Alfred P. Sloan Foundation. For acknowledgments, sources of research support, and disclosure of the author's or authors' material financial relationships, if any, please see http://www.nber.org/chapters /c12684.ack.

1. An exception was the late 1940s when there was a brief concern about engineering surpluses.

technological preeminence. But economists found little evidence of a classic market "shortage" in labor market data on wages, employment, and graduates (Arrow and Capron 1959; Blank and Stigler 1957; Hansen 1967). Ensuing work recognized that the labor market for engineers functions differently from labor markets where demand and supply clear markets quickly because engineering education requires intensive and highly prescribed curriculum, causing supply to lag demand, and that because firms cannot easily find substitutes for engineering skills, wages may be driven up considerably in the short term, producing resultant cobweb-type cycles not found in most labor markets (Freeman 1975, 1976; Ryoo and Rosen 2004).

Still, labor shortage fears have persisted into the twenty-first century (Teitelbaum 2014), perhaps most notably in the National Academy of Science and National Research Council's report, *Rising above the Gathering Storm*. Invoking Churchill's characterization of Germany's threat to Europe in his book, *The Gathering Storm*, the report was written to raise an alarm about the growing threat to the United States by the rise of other nations' science and technology capabilities and insufficient U.S. investment in its own research and development (R&D). Warning that "The nation must prepare with great urgency to preserve its strategic and economic security" (National Academy of Science et al. 2007, 4), the report recommended increasing the number of engineering graduates to keep up with the large numbers graduating in countries such as China. As with the earlier post–World War II cries of shortage and calls for more engineers, these recommendations lacked clear and convincing evidence of unmet domestic demand.[2]

This chapter provides background information on the engineering workforce and trends in the supply of new engineers to the labor market that set the stage for the rest of the book.

The chapter begins with an overview of the engineering workforce and the changes in its detailed occupational composition and distribution across industries over time. Since demand is thought to generally lead supply in this market, we then review the factors influencing the demand for engineers. This includes changes in the demand for engineers across industries, fluctuations in government demand, and replacement demand. This discussion provides context for interpreting trends in the supply of new engineers with undergraduate degrees and graduate degrees from American colleges and universities, which are presented in section 1.3.

The data in the demand section come from the decennial census, the American Community Survey (ACS), the Bureau of Labor Statistics' Occupational Employment Statistics (OES), and the National Science Foundation's (NSF) semiregular Scientists and Engineers Statistical Data System

---

2. While *Gathering Storm* gained the most attention, other reports endorsed the call for more engineers and scientists, most prominently the President's Council on Jobs and Competitiveness ([2011], 33–34; see Salzman [2013], Lynn and Salzman [2010], Lowell and Salzman [2007], and Teitelbaum [2014] for reviews of *Rising above the Gathering Storm* and related reports).

(SESTAT) that combines information from three NSF surveys, the National Survey of College Graduates, the National Survey of Recent College Graduates, and Survey of Doctoral Recipients (SDR) to construct a nationally representative sample of college-degree holders. The chapter focuses on the subset of degree holders in the SESTAT data who earned at least one degree in an engineering field, but it uses ACS and OES data rather than SESTAT for time-series analysis due to changes in the SESTAT sampling frame.[3] The data in the supply section comes from the Integrated Postsecondary Education Data System (IPEDS), which collects detailed annual enrollment and graduation information from the approximately 6,700 postsecondary institutions that accept federal financial aid.

While there is value in analyzing all persons with engineering degrees regardless of where they work, or all persons who report working in an engineering occupation regardless of their education, many of these analyses using the SESTAT and ACS are restricted to persons working as engineers with engineering degrees. These restrictions make the SESTAT analyses more comparable to the IPEDS data. This leaves out engineering graduates who work in other occupations[4] and persons who say they work in an engineering occupation but do not have a bachelor's degree in engineering.[5] The OES data cannot be restricted by field of degree.

1.2   The Engineering Workforce: Demand and Salaries

An engineer is defined by the Bureau of Labor Statistics—across the various subfields—as a worker who designs, develops, and tests solutions to technical and physical problems faced by industry and government. Although popular analyses and policymakers typically refer to engineers as a single group, with the assumption that this represents a more or less homogeneous workforce and labor market, it is instead a remarkably heterogeneous workforce in terms of education, skills, industry representation, and work performed. Civil engineers, for example, are primarily involved in construction work and employed in independent engineering firms and government ("public administration" is the title used in standard industrial classification statistics). In contrast, aerospace engineers are seldom self-

3. In 2010, the SESTAT was drawn from a new sampling frame, the ACS. Prior to 2010 the sampling frame of the SESTAT was the decennial census. Wage and salary data, particularly for engineering graduates working as managers, revealed considerable instability between 2008 and 2010, suggesting that the 2010 and 2013 SESTAT should be interpreted with caution in the context of any time-series comparison with prior years of that survey.

4. In 2013, for example, only 38 percent of all engineering bachelor's degree holders reported they worked as engineers.

5. Between one-fifth and one-quarter of all persons who report working as engineers have college degrees outside engineering and nearly 12 percent have no postsecondary degree at all, presumably obtaining their skills through on-the-job training or working in a highly firm-specific task.

employed or employed by engineering firms, working instead for large aerospace manufacturing firms. Since aerospace engineers are reliant on large military and civilian aviation industry contracts, they are less affected by normal business cycle dynamics than civil engineers whose work is directly affected by levels of construction activity. Mechanical engineers are affected by the business cycle, but their employment levels are sensitive to the decisions of individual firms about the global location of manufacturing operations in a way that might not affect civil engineers.

Although engineering work is a crucial component, if not driver, of much technology innovation, a small portion of the overall engineering workforce is working directly on new technologies or in new product industries. NSF estimates that two-thirds of engineers report research or development as either a primary or secondary work activity (National Science Board 2014, table 3-10), but interpreting the meaning of this statistic for engineering is somewhat different from other occupations. In engineering, a civil engineer might, for example, consider each building project "development" work and it is, in fact, new development in the sense that each new construction project is different from past projects and may require new engineering solutions and innovation. However, this is not what is generally regarded as the type of innovation of interest to policymakers. Most engineers are not creating new technology or developing new products or industrial processes; they are busy designing bridges, roads, power plants, factories, and buildings, and running manufacturing operations. Alternatively, using the industry segment to identify "innovation" work indicates just under 5 percent of engineers (or just over 75,000) are in "scientific research and development services," though many more are involved in key innovation activities in other industries and thus this statistic greatly understates the role of engineers in innovation. Engineering is clearly important to innovation, but most engineers are engaged in "development" rather than the "research" types of R&D that are generally regarded as the innovation of new products. At the same time, the role of engineering outside of formal R&D occupations or work is still vitally important to the use of new products and improvements in production processes, a vital component of commercialization of innovation (see Helper and Kuan, chapter 6, this volume; Barth, Davis, Freeman, and Wang, chapter 5, this volume). Thus, when discussing the role of engineers in innovation it is important to specify the particular subfields, occupational roles, and types of innovation that are of interest rather than refer to the entire engineering fields of work or occupations.

The employment patterns of engineers vary according to the specific field and industry of employment as different segments of the American economy, such as construction or manufacturing change. Table 1.1 presents selected shares of engineering occupations employed in major industry categories in 1980 and 2010 for each detailed engineering occupation. Some occupations, like computer engineering and environmental engineering,

Table 1.1   Engineering employment in industry: Percentage of top three industries (1980 or 2010) for each engineering occupational subfield

Occupation	Largest industry (1980 or 2010)	1980	2010	Second-largest industry	1980	2010	Third-largest industry	1980	2010
Aerospace	Manufacturing	88	86	Professional, scientific, and technical services	4.1	7	Public administration	5	3.8
Biomedical & agricultural	Manufacturing	53	54	Education, health, arts, entertainment, and other services	24	29	Public administration	14	3.3
Chemical	Manufacturing	81	90	Professional, scientific, and technical services	9	6			
Civil	Construction	40	24	Professional, scientific, and technical services	24	60	Public administration	16	8
Computer	Professional, scientific, and technical services	—	37	Manufacturing	—	32	Finance, insurance, and real estate	—	9
Electrical	Manufacturing	50	45	Professional, scientific, and technical services	11	21	Information technology	16	11
Environmental	Professional, scientific, and technical services	—	43	Public administration	—	19	Transportation and utilities	—	15
Industrial	Industrial	80	76	Professional, scientific, and technical services	3.5	12			
Marine	Transportation and utilities	58	20	Manufacturing	21	48	Professional, scientific, and technical services	9	15
Materials	Manufacturing	78	90	Professional, scientific, and technical services	7	5			
Mechanical	Manufacturing	76	73	Professional, scientific, and technical services	10	14			
Mining & petroleum	Mining and extraction	80	72	Manufacturing	6	10	Professional, scientific, and technical services	5	8
All other engineers	Manufacturing	37	31	Professional, scientific, and technical services	35	36	Public administration	12	15
Engineering technicians	Manufacturing	47	42	Professional, scientific, and technical Services	12	13	Education, health, arts, entertainment, and other services	10	10

were not recorded as separate fields in 1980. Presumably the small number of engineers doing comparable work in 1980 was categorized as electrical engineers (computer) or civil engineers (environmental).

Engineers are concentrated in the manufacturing sector, with half working in that industry in 1980, declining to 44 percent in 2010. Several engineering fields are heavily concentrated in manufacturing such as aerospace, chemical, industrial, marine,[6] and materials engineers, in addition to engineering technicians. Biomedical and agricultural engineers also have been employed primarily in manufacturing, although roughly one-quarter of this field is also employed in the health sector, with the health-sector employment presumably made up primarily of biomedical engineers. The manufacturing sector shed well over one-quarter of its total workforce between 1980 and 2010, mostly from 2000 to 2010, due to a combination of productivity improvements, offshoring, and loss of market share to foreign competition (Congressional Budget Office 2008). Despite this substantial decline in the total manufacturing workforce, the number of engineers employed in manufacturing was virtually identical in 1980 and 2010 (1,071,700 in 1980 compared to 1,077,400 in 2010). Over the same period, manufacturing production increased by almost 80 percent with a smaller total workforce and a stable number of engineers employed.[7]

The share of engineers in the professional, scientific, and technical services (PSTS) industry grew from 14 percent to 25 percent over the past thirty years. In absolute terms, the number of engineers in this sector more than doubled from 301,800 in 1980 to 614,700 in 2010. The PSTS industry is composed of engineering consulting firms, so their growth is likely due to the outsourcing of engineering tasks from in-house engineering to consulting engineering firms (Yuskavage, Strassner, and Medeiros 2009). The largest shifts of engineering occupations into the PSTS industry have been civil and electrical engineers, also two of the largest engineering occupations, comprising almost 18 percent and over 13 percent, respectively, of all engineers.[8] Electrical engineering concentration in the manufacturing sector fell between 1980 and 2010, while civil engineering concentration declined in the construction and "public administration" (government) sectors, from 40 percent to 24 percent, and 16 percent to 8 percent, respectively; civil engineers in PSTS grew from 24 percent to 70 percent of all civil engineering employment. This

6. Caution should be used in interpreting percentage changes in occupations with small workforces such as marine engineering (in 2010 there were 11,000 marine engineers in the United States). Small sample sizes, the change from using the census long form to the ACS sample, and changes in the industry and occupational classifications could account for instability in the population estimates.

7. Authors' calculations from the Federal Reserve Board's G.17 Industrial Production and Capacity Utilization tables (Board of Governors of the Federal Reserve System 2011).

8. Authors' calculations from the 2010 census. The single largest engineering occupational group, a miscellaneous engineering category, comprises almost a quarter of all engineers.

Table 1.2   **Industrial sectors employing engineers**

Sector	Number	Percent
Agriculture and forestry	1,931	0.1
Mining	34,064	1.8
Construction	96,321	4.9
Manufacturing	868,505	45
Wholesale	15,828	0.8
Retail	13,215	0.7
Transportation and warehousing	26,158	1.3
Utilities	81,769	4.2
Information	57,830	3.0
Finance, insurance, and real estate	16,559	0.9
Prof., sci., and tech. services	548,406	28
Educational and health services	36,332	1.9
Arts and entertainment	5,236	0.3
Other services	7,718	0.4
Public administration	140,448	7
Total	1,950,320	100

Source: Authors' calculations from the 2010 ACS.

sector shift out of manufacturing, construction, and public administration may overstate the shift in engineering activity—of independent engineering firms in the PSTS sector, many may still be providing engineering services to a similar profile of client industries (e.g., construction, manufacturing). The PSTS industrial sector includes architectural services as an industrial subsector, which is particularly important in accounting for the growth in employment of civil engineers in PSTS. (See table 1.2.)

Engineering graduates working as engineers are paid well.[9] Table 1.3 presents the real (2014 dollars) median salaries for all engineering fields in selected years between 2005 and 2014. Petroleum and computer hardware engineering are among the highest paid fields (over $100,000 in 2014), with civil engineering a consistently lower paid field (in the mid-$80,000 range during this period).

The different pay ranges and changes over time for selected engineering fields are shown in figure 1.1. Civil, industrial, and mechanical are lower-paid fields and show minimal increase in median wages over the decade, with no wage growth in the last period, 2011 to 2014. Petroleum engineer stands out as both a substantially higher-paying field and as the one occupation that experienced sharp increases in earnings (see Lynn, Salzman,

9. The median earnings of engineers in 2014 dollars presented in table 1.3 are all substantially above the annualized median earnings of all management and professional workers in 2014 reported by the Bureau of Labor Statistics (2015), where median weekly earnings for full-time workers are annualized by multiplying by fifty-two weeks.

Table 1.3   Engineering graduate real (2014 dollars) median salaries by occupation

Occupation	2005 ($)	2008 ($)	2011 ($)	2014 ($)
Petroleum	118,004	131,000	146,269	147,520
Computer hardware	105,665	110,153	106,676	110,650
Aerospace	103,580	103,336	109,317	107,700
Nuclear	109,931	109,680	110,675	104,630
Chemical	96,040	97,596	104,655	103,590
Mining and geological incl. mining safety	90,997	87,865	94,794	100,970
Electronics except computer	96,961	97,497	99,635	99,660
Marine and naval architects	90,088	85,677	96,541	99,160
Engineers, all other	94,028	97,948	97,099	96,350
Electrical	92,197	93,846	93,878	95,780
Biomedical	91,373	89,195	92,994	91,760
Materials	86,537	92,582	91,342	91,150
Mechanical	84,852	85,985	87,932	87,140
Civil	84,221	86,380	87,048	87,130
Environmental	85,724	85,732	87,711	86,340
Industrial	83,033	83,280	84,027	85,110
Health and safety	81,506	81,180	82,659	84,850
Agricultural	80,451	80,102	82,512	75,440
All engineers	90,518	91,978	93,659	93,626

Source: Authors' calculations from 2014, 2011, 2008, and 2005 OES.

and Kuehn, chapter 8, this volume).[10] Chemical engineering also showed increases, likely due to a small but significant number of chemical engineers working in the petroleum industry and paid substantially higher wages than other chemical engineers. Of note is that, with only a few exceptions, median earnings levels have shown little wage growth over the past decade. The wage levels would suggest that only in a few fields has there been growing demand for engineers.

From 1993 to 2013, no more than about half of workers with a degree in engineering actually reported that they were working as engineers, as compared to about one-third of all science, technology, engineering, and mathematics (STEM)-degree labor force participants who work in a STEM occupation (exclusive of social sciences where bachelor's degrees rarely confer professional status).[11] Of recent graduates, about two-thirds of those with four-year engineering degrees report they are in an engineering or other STEM occupation, as compared to only about a half of all STEM graduates who enter a STEM occupation (Salzman, Kuehn, and Lowell 2013; Salzman

10. See Lynn, Salzman, and Kuehn (chapter 8, this volume), for analysis of the changes in this occupation.
11. Since the wages in the previous tables and figure are only for those engineers working in an engineering occupation, they relate to only about half of all engineers in the workforce.

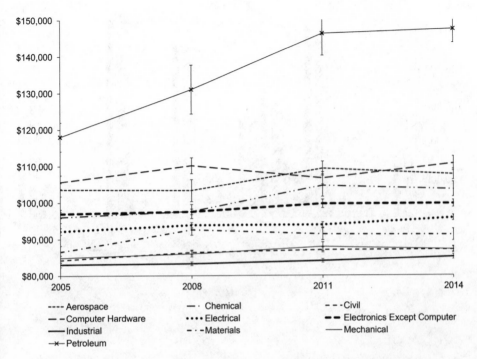

Fig. 1.1   Real earnings (2014 dollars) of engineers by selected fields, 2005–2014
Source: Authors' calculations from 2014, 2011, 2008, and 2005 OES.

2015).[12] This suggests that although many engineering graduates do not enter an engineering career, those who do persist in a STEM career at higher rates than graduates in other nonsocial science STEM fields. However, the employment prospects of engineers vary significantly by field.

The workforce size and changes over the past decade are shown in figure 1.2. Employment growth occurred in petroleum (Lynn, Salzman, and Kuehn, chapter 8, this volume), biomedical, mechanical, industrial, and electrical engineering fields. Aerospace and civil engineering changes reflect the cyclical changes in their respective fields, with civil engineering reflecting the construction expansion and then collapse, followed by a recovery at the end of the past decade.

1.2.1   Government Demand for Engineers

Government is a large employer of engineers, both directly as public employees (7.4 percent of all engineers in 2010 as reported in the OES statis-

---

12. Authors' tabulation of ACS data reported in U.S. Census Bureau (2012) and Salzman (2015).

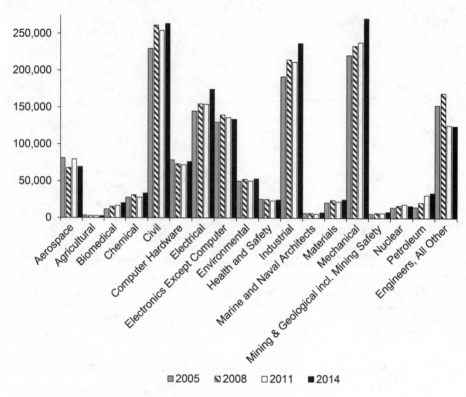

Fig. 1.2   Employment of engineers by occupation, 2005–2014
Source: Authors' calculations from 2014, 2011, 2008, and 2005 OES.

tics) and indirectly in private companies working on government contracts. In the post–World War II period, the federal government expanded its role as a significant funder of public infrastructure and research and development, maintaining a large standing military and expanding the role of science and engineering in defense strategy, from weapons development to espionage. Notable cases of sharp increases in the public demand for engineering labor that had a dramatic effect on the engineering labor market range from the interstate highway system to increased R&D for new military systems. Major defense investments began in the late 1970s in the Carter administration through the mid-1980s, when it appears military R&D slowed and defense-related engineering employment declined during the latter half of the Reagan administration.[13] Federal policymakers have thus viewed engi-

---

13. As discussed in section 1.3, military R&D that affects engineering employment may occur in a different cycle from overall military spending, preceding overall military budget increases by a number of years, and may not be paid for by the military until years after the defense

neers (and scientists) as essential to the military and economic strength of the nation.

Following World War II, policymakers have focused on programs that could develop the science and engineering workforce as a particular goal of more general government efforts to improve the education of the U.S. population. In doing so, federal policymakers shifted their approach to engineering labor supply from "markets-to-manpower" planning (Kaiser 2002). More generalized educational policies such as the GI Bill and the expansion of university funding also led to increasing numbers of students pursuing engineering as an occupational route off the farms and into the middle class. Manpower studies assessing the adequacy of the science and engineering workforce became commonplace across a variety of agencies, using more sophisticated modeling and forecast methods over time (McPherson 1986). Vannevar Bush's 1945 report *Science, the Endless Frontier* developed the vision for postwar government science policy that still influences policy today. Bush's discussions of the science and engineering workforce focused on the production of new scientists and engineers to support the expansion of science and engineering for the national defense and advancement of discovery and innovation. In 1953, the new National Science Foundation took on primary responsibility for monitoring the science and engineering workforce and the maintenance of a National Register of Scientific and Technical Personnel (National Science Foundation 1953).

The federal government developed a direct interest as an employer in the supply of engineers, as well as a more general policy interest in ensuring an adequate supply for industry in general, and military contractors in particular. Government commissions and agencies regularly assessed the

---

contractor incurs the expense, including the associated increase in employment. Fox (2011, 22), in his in-depth history of military procurement, notes that:

> There is often an earlier informal acquisition process that has its origin in defense laboratories or defense contractor firms, where engineers conceive of a new device or a new subsystem. Representatives of a firm may approach a military service, describe how they believe a device or subsystem will enhance the defense capability of the service, and then help the service prepare the justification and RFP to conduct a more formal study of the idea. This assistance nurtures the idea until it evolves into a military requirement.
>
> A defense firm wishing to obtain a contract to develop a new weapon system usually becomes involved in the program two to four years before a formal RFP is issued, or it is unlikely to qualify as a prospective contractor. This involvement generally means assisting the buying service in defining elements of the planned weapon system. The cost of conducting this initial work generally becomes part of contractors' overhead costs (for example, bid and proposal expense or independent research and development expense), which the Defense Department usually reimburses in part or in full.

The first phase of the military buildup during the early 1980s of the Reagan administration was accompanied by many reports of, and congressional investigations into, "fraud, waste, mismanagement and abuse" and various attempts to reform the procurement process (Fox 2011). It may be that the increased attention to military contractor practices also curtailed contractor R&D by the mid-1980s.

engineering workforce supply, and the resulting reports ranged from alarm at impending labor shortages[14] to empirical assessments that found supply to be adequate, fluctuating with cyclical changes in demand. In 1986 a report by the National Academy of Engineering on the effect of increased Reagan-era defense spending on the engineering labor market concluded that "evidence drawn from a variety of sources does not suggest pervasive or serious industrial shortages" (National Research Council 1986). The most recent report from the National Research Council (2012) on the STEM workforce supply for the Department of Defense, led by the same committee chairman of the earlier 2007 National Research Council report about impending shortages and crises in STEM, found the engineering, scientific, and technical workforce supply to be sufficient with the exception of cybersecurity experts, anthropologists, and linguists, for which the Department of Defense had unmet demand and difficulty recruiting.

The federal government is the primary source of government demand for weapons systems, biomedical and other research, and large infrastructure projects. State and local governments differ both in the scale and the nature of their engineering activities. These governments almost exclusively employ or contract with civil or environmental engineers for infrastructure projects and regulatory services. The skill sets of these engineers are generally more homogeneous, and they are more likely to work in smaller firms than in larger specialized firms that work on federal contracts. Table 1.4 presents the composition of the federal engineering workforce in 2010 and the federal government's share of the national engineering workforce by detailed occupational categories. Table 1.5 presents the same information for state and local governments.[15] Federally employed aerospace engineers make up less than 10 percent of the engineers directly employed by the federal government. A quarter of federally employed engineers have no specialized field identified at all. A similar share are electrical and electronics engineers (4.4 and 18 percent, respectively), with smaller proportions working as mechanical engineers (12 percent) and civil engineers (11 percent). In selected detailed occupations, the federal government employs a nontrivial

14. For example, the 1962 Gilliand panel reported that, "Impending shortages of talented, highly trained scientists and engineers threaten the successful fulfillment of vital national commitments. Unless remedial action is taken promptly, future needs for superior engineers, mathematicians, and physical scientists seriously outstrip supply" (President's Science Advisory Committee 1962, 1). A recent example is the National Academy's (2007, 3) *Rising above the Gathering Storm* report, which was "deeply concerned that the scientific and technological building blocks critical to our economic leadership are eroding at a time when many other nations are gathering strength."

15. The data is from the 2010 OES. Federal employment of engineers in the OES data from 2008 is of comparable magnitude to NSF estimates using data from the U.S. Office of Personnel Management (National Science Foundation 2008). Total public employment of engineers from the OES in 2006 is somewhat higher than from SESTAT (166,630 in the OES and 144,250 in SESTAT), potentially because of differences between self-reported occupational categories and occupations reported by employers. The SESTAT is also restricted to college degree holders, and a small share of engineers does not hold four-year college degrees.

Table 1.4   **Engineering occupations employed by federal governments in 2010**

Engineering occupation	Number	Percent of federal engineers in detailed occupation groups	Percent of engineers nationally working for federal government
Aerospace engineers	9,220	9	12
Agricultural engineers	400	0.4	16
Biomedical engineers	500	0.5	3.3
Chemical engineers	1,130	1.2	3.9
Civil engineers	10,630	11	4.3
Computer hardware engineers	4,430	5	7
Electrical engineers	4,260	4	2.9
Electronics engineers, except computer	17,790	18	13
Engineers, all other	25,450	26	18
Environmental engineers	3,800	3.9	8
Health and safety engineers, except mining	710	0.7	3.0
Industrial engineers	1,340	1.4	0.7
Marine engineers and naval architects	910	0.9	16
Materials engineers	1,320	1.4	6
Mechanical engineers	11,710	12.1	5
Mining and geological engineers	160	0.2	2.6
Nuclear engineers	2,730	2.8	15
Petroleum engineers	330	0.3	1.2
Total	96,820	100	7

Source: Authors' calculations from the 2010 OES.

Table 1.5   **Engineering occupations employed by state and local governments in 2010**

Engineering occupation	Number	Percent of state and local engineers in detailed occupation groups	Percent of engineers nationally working for state and local government
Aerospace engineers	0	0.0	0.0
Agricultural engineers	0	0.0	0.0
Biomedical engineers	40	0.0	0.3
Chemical engineers	110	0.1	0.4
Civil engineers	61,030	69	25
Computer hardware engineers	120	0.1	0.2
Electrical engineers	3,130	3.5	2.1
Electronics engineers, except computer	540	0.6	0.4
Engineers, all other	6,570	7	4.7
Environmental engineers	11,120	13	22
Health and safety engineers, except mining	2,940	3.3	13
Industrial engineers	510	0.6	0.3
Marine engineers and naval architects	70	0.1	1.2
Materials engineers	300	0.3	1.4
Mechanical engineers	1,950	2.2	0.8
Mining and geological engineers	410	0.5	7
Nuclear engineers	70	0.1	0.4
Petroleum engineers	0	0.0	0.0
Total	88,910	100	6

Source: Authors' calculations from the 2010 OES.

share of all engineers. For example, although less than 1 percent of federal engineers are marine engineers, the estimated 910 federal workers make up almost 16 percent of that engineering occupation nationally.

The state and local engineering workforce is far less occupationally diverse. Sixty-nine percent are civil engineers and 13 percent are environmental engineers. State and local governments employ almost one-quarter of civil engineers and over one-fifth of environmental engineers, nationally.

The most detailed information on direct federal employment of scientists and engineers comes from internal personnel data provided by the U.S. Office of Personnel Management and the Defense Manpower Data Center and published on an irregular basis by NSF. These data suggest that federal engineering employment after the Cold War peaked at just under 98,000 in 1992, dropping over the course of the decade to a low of just over 82,600 in 2000. After 2000, federal engineering employment grew steadily again, reaching a peak of just over 87,000 in 2004. NSF has not published updates on these federal employment figures for years more recent than 2005. The distribution of these changes are relatively even across fields; no field contributed an especially disproportionate amount to the change.[16]

The share of engineering graduates working in an engineering occupation that is either employed by the government or working on a government contract or grant has remained relatively steady during the post–Cold War period, ranging from around 36 to 39 percent of the total (table 1.6), with most of those engineers working on government contracts rather than directly for the government. The share of these engineers working on government contracts (rather than as direct employees) has increased steadily over time from 60 percent in 1993 to 71 percent in 2013. These engineers may not be exclusively doing government work, but may also have private contracts; to the extent that engineers working as contractors are not exclusively working as full-time equivalent government contractors, the employment shift could be overstating the full-time equivalent engineering employment on federal work (either on a contract or grant or in direct federal employment).

The government is likely to continue to play an important role in the demand for engineers, particularly in specialized occupations like aerospace through defense purchases, nuclear engineering directly through defense contracts and indirectly through roles in regulation and any issuing of new reactor permits, and civil engineering through public works projects, particularly at the state and local level. Funding levels for the National Institutes of Health, alongside growth of medical devices firms, will be an important force shaping the demand for biomedical engineers inside and outside of government. The extent of continued demand for civil and environmental engineers by state and local governments is likely to be highly contingent on government infrastructure investments as well as private construction.

16. See the National Science Foundation's (1995, 2005, 2008) reports on federal scientists and engineers.

Table 1.6   Engineering graduates working as engineers and on government work
                   (direct or on contract/grant)

	(1) Percent employed by the government or on government contract (%)	(2) Number employed by the government or on government contract	(3) Contract share of government engineers (share of column [1]) (%)	(4) Number employed on government contract
1993	38.07	382,196	60	227,799
1995	36.87	382,489	62	238,948
1997	34.75	374,507	63	237,522
1999	—	—	—	—
2003	36.27	410,683	68	277,857
2006	36.80	454,798	70	316,147
2008	37.14	458,097	70	321,624
2010	38.29	432,805	70	302,650
2013	39.82	456,965	71	326,311

Source: Authors' calculations from SESTAT 1993–2013. The sample is restricted to employed respondents who earned a bachelor's degree or higher in engineering and earned their highest degree in the United States, and is weighted to be nationally representative. The 1999 SESTAT is excluded because it does not specify whether respondents work on government contracts. The SESTAT asks survey respondents, "was any of your work during [year] supported by contracts or grants from the U.S. government?" The SESTAT engineer population is 1,147,000, compared to the ACS engineer population of 1,950,000, due to differences in sampling and population coverage; the ACS population is all who self-report work in an engineering occupation, regardless of degree level or where the degree was awarded.

1.2.2 Retirement and Replacement Demand

A concern about the engineering labor market is whether large worker attrition due to retirement and older workers leaving the field can be filled. Even in engineering fields that are not growing, replacement demand may be a major source of demand for new graduates. The impending retirement of baby boomers has led to concern that not enough graduates are being produced to replace retirees and that transferring knowledge from more experienced to less experienced engineers will pose problems for engineering continuity (Gibson et al. 2003). Freeman (2007) provides an empirical framework for investigating this issue and finds that "demographic changes have not historically been consistently associated with changes in labor market conditions, even for the young workers whose position is most sensitive to changing market realities." Freeman (2007, 5) assesses replacement demand for all occupations and we use his empirical framework to analyze the engineering labor market specifically. The analysis finds evidence that demand for younger engineers is not related to aging in the engineering workforce. This result is somewhat weaker than Freeman's analysis of all occupations, which found a negative relationship between the proportion of workers over age fifty-five in an occupation and growth of younger cohorts in those fields. He concludes that older workers are concentrated in occu-

Table 1.7   Estimated relationships between occupational employment of engineering graduates older than fifty-five in 2003 and employment of other age groups

	Age group in 2013		
	25–34	35–44	45–54
Share of workers age fifty-five and older, 2003	0.085	0.035	0.366***
	(0.155)	(0.066)	(0.085)
Share of workers in specified group, 2003	0.857***	0.327***	0.333***
	(0.149)	(0.082)	(0.126)
Constant	0.093*	0.126***	0.102***
	(0.055)	(0.036)	(0.036)
R^2	0.408	0.142	0.190
Observations	118	118	118

Source: Authors' calculations from SESTAT 2003 and 2013. The sample is restricted to employed respondents who earned a bachelor's degree or higher in engineering and earned their highest degree in the United States, and is weighted to be nationally representative.
***Significant at the 1 percent level.
*Significant at the 10 percent level.

pations with declining employment, and thus not replaced; employment demand for younger workers is, instead, predominately in growing occupations. While our results do not suggest that older workers are concentrated in labor markets that are in decline, they also do not suggest a major role for replacement demand in the market for new engineers.

Occupation-sector-educational-attainment cells serve as the unit of analysis in the regressions in table 1.7, which estimate the employment share of various age groups in 2013 on that specified group's employment share a decade earlier in 2003 and the share age fifty-five and older in 2003. For example, one cell is "electrical engineers in the private sector with master's degrees," while another is "electrical engineers in the public sector with master's degrees." Occupation-sector-education cells with fewer than twenty observations in the 2013 SESTAT were omitted from the analysis to ensure the reliability of the estimates of employment shares. One hundred and eighteen occupation-sector-education cells were available in both 2003 and 2013. As with other analyses using the SESTAT data, the sample is restricted to holders of engineering degrees who earned their highest degree in the United States. Occupations that are in equilibrium, where new labor supply is sufficient for meeting replacement demand, would have employment shares for each age group in 2013 that are comparable to the employment shares of that age group in 2003. We would expect such an equilibrium to have significant coefficients for their specified age group, and no significant relationship between the employment share of engineers that are age fifty-five or older. In an occupation that is experiencing the retirement of a large population of older workers, and responding with a major recruitment of younger workers, we would expect a positive association between the employment share of engineers that are fifty-five or older in 2003 and the worker share of a younger age group in 2013.

The regressions provide no evidence that the share of younger workers (ages twenty-five to thirty-five) increases in 2013 when a larger share of the occupation-sector-education-group's workers are over the age of fifty-five in 2003. The regressions show a larger positive and statistically significant relationship between the share of workers older than fifty-five in 2003 and the share of workers age forty-five to fifty-five in 2013. This finding is consistent with a declining occupation group rather than an occupation making a major effort to recruit young workers to replace retiring workers. As older workers in a declining occupation retire and few younger workers are hired because there is no expectation of a future expansion of the occupation, the remaining middle-aged workers represent an increasing share of the workforce.

The finding that in general retiree replacement demand is not an important source of demand for new engineers does not imply that in all cases replacement demand is negligible. Petroleum engineering provides a recent example of an occupation with an aging workforce that, in combination with new industry growth due to technology innovation and product demand, until recent years was growing rapidly and drew many new engineering graduates into the field with the incentive of higher salaries. Analysis of the petroleum engineering industry is discussed in more detail by Lynn, Salzman, and Kuehn (chapter 8, this volume).

1.3 Engineering Education and the Supply of New Engineers

Engineers require specialized training, predominately at the undergraduate level, though in some cases at the graduate level. As with most specialized professions, nonengineers cannot be easily substituted for engineers to meet demand. While conceivably former engineers can move back into the occupation from other occupations, this sort of labor "backtracking" is relatively rare (Biddle and Roberts 1994). Immigrants trained abroad can also add to the stock of engineers, but the primary source of new engineering labor is through the entrance of new graduates into the labor market (some of whom may be immigrants themselves). This section presents trends in the supply of new engineering graduates in the aggregate and by field of engineering. New engineer supply trends are tracked using data from the Integrated Postsecondary Education Data System (IPEDS), which collects detailed annual enrollment and graduation information on all postsecondary institutions that accept federal financial aid (approximately 6,700 institutions). Data come from engineering degree programs and specifically exclude computer science graduates and "engineering tech-

nology" programs that primarily produce software engineers and technicians, respectively, rather than engineers. The IPEDS data are made publi-
cally available from 1980 to the present in detailed institution-level files and
from 1966 to the present at a more aggregated level.[17]

1.3.1   Bachelor's Degree Trends

Figure 1.3 presents data on engineering graduates from 1966 to 2013 by
the level of degree awarded. Bachelor's degrees are measured according to
the scale on the left axis, and range between 60,000 and 90,000 degrees
awarded every year during the past thirty years since the mid-1980s, after
almost doubling between 1976 and 1985. In this figure, graduate and associate's degrees are shown on the right axis, reflecting the much smaller number of graduates at those degree levels. Master's degree awards have been
increasing steadily since 1966, with rapid growth late in the first decade of
the twenty-first century. Almost 45,000 master's degrees in engineering were
awarded in 2013, triple the number of awards that were typically awarded
annually between 1966 and 1980. Considerably fewer doctoral and associate's degrees have been produced annually during this period, with no dis-
cernable increase in these awards over time.

There are five distinct periods in the bachelor's degree award trend in figure 1.3. First, between 1966 and 1976, bachelor's degree awards fluctuated
between 40,000 to 50,000 annually, ending at the same level as the beginning
of this ten-year period. Then, between 1976 and 1985, bachelor's degrees in
engineering increased from under 40,000 to almost 78,000. This period of
growth was followed by a decline in awards through the late 1980s to somewhat more than 60,000 bachelor's degrees produced annually by 1990. After
1990, engineering awards stabilized at a level slightly higher than 60,000
awards per year, and maintained this level through the dot-com bubble in
the late 1990s, until approximately 2001.

Since 2001, the number of bachelor's degrees in engineering awarded
annually have steadily increased, reaching the level of degrees awarded in
1985 (roughly 70,000 per year) in 2008 and then surpassing it in 2011; by
2013, almost 89,000 bachelor's degrees in engineering were awarded. Graduate degrees in engineering have experienced more consistent growth since
1980 than bachelor's degrees. Between 1981 and 2004, a period when there
were almost no net gains in bachelor's degrees awarded, the number of master's degrees awarded in engineering doubled from 16,000 to nearly 34,000.
Doctorates more than doubled from 2,500 to almost 6,000 over this period.

Each of these degree levels can be decomposed into more detailed engineering fields in which degrees are awarded. As discussed earlier, "engi-

---

17. Engineering fields and programs in this section are identified using the IPEDS Classification of Instructional Programs, or CIP codes. These codes may not reflect organizational
structure of the departments themselves in all cases.

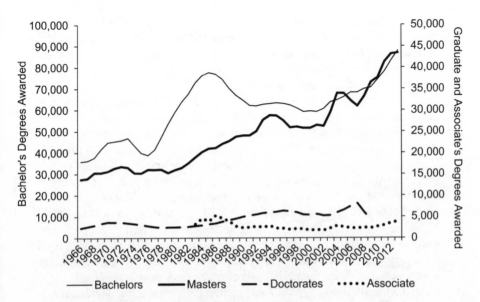

Fig. 1.3   Engineering graduates, 1966–2013

Source: Authors' calculations from engineering degrees in the IPEDS, 1966–2013.

Note: Bachelor's degrees are shown on the left axis; doctoral, master's, and associate's degrees are shown on the right axis.

neering" is an aggregate occupational category that comprises a number of distinct fields that may have some common educational requirements but are occupationally quite different. An electrical engineer, for example, may be designing semiconductors while a civil engineer is designing buildings, with little commonality in skills, technology, or knowledge content, and working in industries that may have very different business and employment dynamics. Analysis of supply and demand thus requires a more disaggregated analysis of the major engineering fields.

In this section we examine the trends in each of the major engineering occupations identifiable back to 1966: electrical, which is a field largely supplying the manufacturing sector and the information technology (IT) and computer industries; mechanical, which is also predominately in manufacturing, concentrating on the design and maintenance of manufacturing equipment and systems; civil, which supplies the construction sector and the independent professional-services sector (e.g., architectural firms and engineering consulting firms) and local governments; and chemical, industrial, aerospace, and materials engineering, which are all concentrated in manufacturing. All other engineering fields are grouped in an "other" category, which includes a range of specialty fields from petroleum engineering to computer engineering and biomedical engineering. We also look at detailed occupations that form a relatively small portion of the total but are important growth sectors: biomedical engineering, which has grown more recently and appears to be a route both into engineering (e.g., medical devices) and, to some extent, medical school; and computer engineering, which grew rapidly over the late 1990s but has since experienced a decline. Together, engineering degrees in these six fields (electrical, civil, chemical, mechanical, computer, biomedical) accounted for 77 percent of all engineering bachelor's degrees awarded in 2008.

Trends in bachelor's degrees awarded in these six fields are presented in figure 1.4. A major driver of the increase in engineering graduates in the mid-1980s was the high production of electrical engineering degrees, which made up a third of all engineering bachelor's degrees in that decade. Between 1976 and 1987, the number of electrical engineering bachelor's degrees awarded increased by 270 percent, from just under 10,000 annually in 1976 to almost 27,000 in 1987. Annual electrical engineering awards went into a period of sharp decline after 1987, accounting for a large share of the decline in total engineering degree awards during this period. Between 1987 and 1996, the number of electrical engineering bachelor's degrees awarded annually in the United States fell by over 10,000. Mechanical engineering degree awards track electrical engineering degree awards, but to a more modest extent. Mechanical engineering grew rapidly in the late 1970s and early 1980s, and then it moderated for the next several decades. The decline in mechanical engineering awards after its peak in the mid-1980s from 17,200 in 1985 to 14,263 in 1991 was not as steep as the decline in electrical engineering.

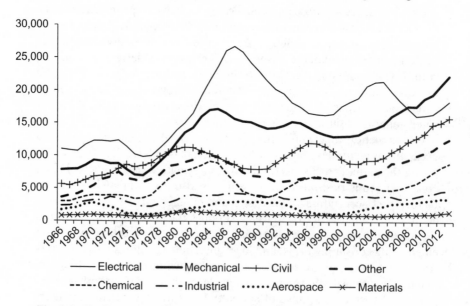

Fig. 1.4   Engineering bachelor's degree awards, 1966–2013
Source: Authors' calculations from engineering degrees in the IPEDS, 1966–2013.

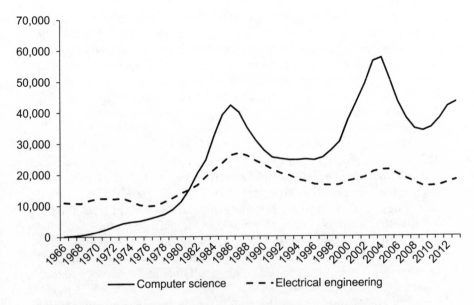

Fig. 1.5   Computer science and electrical engineering bachelor's degrees, 1966–2013

Source: Authors' calculations from computer science and engineering degrees in the IPEDS, 1966–2013.

The increase in these engineering degrees was countercyclical during this period: gross domestic product (GDP) growth declined beginning around 1977 until the early 1980s and overall unemployment increased, reflecting the downward trajectory of rates of GDP growth. Only in the late 1980s does the number of engineering graduates decline along with rising unemployment and declining GDP growth rates. Between 1970 and 1986, bachelor's degrees in computer science increased by almost thirtyfold from 1,500 to 42,200 awards, while over the same period electrical engineering degrees doubled from 12,300 awards to 26,100 (see figure 1.5). Graduates in electrical engineering increased significantly during this period relative to other engineering fields, but its growth and size were eclipsed by those of computer science. Between 1987 and 1996, the number of electrical engineering bachelor's degrees awarded annually in the United States fell by over 10,000, accounting for the major share of overall engineering degree declines. Computer science also experienced a decline in bachelor's degrees awarded during the late 1980s and early 1990s, as well as the early twenty-first century (see figure 1.5), falling from a peak of 42,200 in 1986 to 24,600 in 1994.[18]

The strong growth in these engineering fields while the overall economy

18. Authors' calculations of degrees from the IPEDS.

was in decline during the 1970s and early 1980s appears to be due largely to the growth and change in military technology development and in the semiconductor industry and secondarily by the growing personal computer industry, though the size of its workforce was small during this period. The early to mid-1970s was a period of economic stagnation and decline in demand for engineering. Defense spending declined beginning in the early 1970s, and continued its decline following the complete withdrawal from the Vietnam War in 1975 and the overall economic stagnation following the 1973 Organization of Petroleum Exporting Countries' oil embargo and the 1973–1975 recession. Defense spending declined by 30 percent (in constant dollars) from 1968 to 1973, with a further 10 percent decline by 1976, ending that ten-year period at 60 percent of 1968 spending levels. This occurred alongside civilian aircraft sales declines of nearly 25 percent from 1968 to 1971, with overall job losses of 900,000 in the defense sector from 1969 to 1971, and job losses of 580,000 in the aerospace industries from 1968 to 1972; by 1973, employment in the space program was left at one-third the size of its 1965 workforce (U.S. Congress, Office of Technology Assessment 1992, 106).

In the late 1970s through the mid-1980s, the number of engineering graduates increases in an apparent countercyclical pattern to the overall economy, but it does follow the growth in military technology development, electronics, and computer industries. Although engineering graduation rates reflect changes in military spending on R&D, they do not closely reflect the trajectory of *overall* military expenditures because engineering employment for military technology can follow a different pattern from that for overall defense spending. Technology development will begin before actual production of new systems (sometimes even before the military programs are fully funded) and the involved engineers and other developers may be laid off once the systems go into production, long before production and overall spending declines, or when systems are cancelled during development (U.S. Congress, Office of Technology Assessment 1992). Engineering employment, and electronics engineer and computer scientist employment in particular, for the two decades following the late 1970s, appears to be driven largely by the growth of the semiconductor industry; in addition to demand for semiconductors, development of other electronic technologies using semiconductors, and especially the development trajectory of military systems, as well as changes in the fortunes of the semiconductor industry during this period, drove the engineering and computer science employment cycles.

During the late 1960s through the early to mid-1980s, civilian industries were not expanding their engineering workforces and many industries, especially aerospace, were in severe decline, as noted above. However, even while overall military spending was in decline and then stabilized just before the large Reagan-era Star Wars military program (Strategic Defense Initiative [SDI]) was established in 1984, new technologies were being developed by the military. The systems developed beginning in the late 1970s represented a significant change in technology—they were systems that shifted to more electronics, "black-box" maintenance designs (in which components were swapped out to be repaired off the battlefield), and generally required much greater levels of technical support and maintenance than older systems, alongside a shift to using more nonmilitary personnel contractors to support military missions. The fundamental change to greater electronics-based technology was reflected in an Air Force general's quip: "In the past, the Air Force used to buy airplanes and add electronics. Today the Air Force buys computers and puts wings on them." Forty percent or more of the funds for Department of Defense aircraft are spent on electronics equipment (Fox 2011).

The expansion of military systems beginning in the late 1970s also increased in anticipation of greater military spending in the 1980s, "long before Reagan surprised most of the nation with his ode to Star Wars" (Hiatt and Atkinson 1985). Firms involved in military technology development "had been pushing for new 'defensive' systems prior to their becoming politically popular in the 1980s. Indeed, the big aerospace firms already had done substantial SDI-related work in the 1970s" (DiFilippo 1990). The genesis of SDI was in part the outcome of laser technology development in the 1970s, which was heavily funded by the Department of Defense, spending $100 million a year by 1981 (DiFilippo 1990, 114). It was laser technology that convinced the hydrogen bomb developer Edward Teller and General Daniel Graham, "who identifies himself as the 'midwife' of Star Wars" (DiFilippo 1990, 114), that ballistic missile defense was technically possible. (Teller became director of a private laser technology firm that relied on Department of Defense funding as well as becoming a member of the White House Science Council where, in 1982, he began proposing SDI development to President Reagan [DiFilippo 1990, 115].) In 1981 Rockwell International wrote in its company brochure "Space defense systems will be developed in the near future" and "companies unleashed their engineers in a hunt for new weapons, new technology" (Hiatt and Atkinson 1985). And this was true throughout the branches, as two journalists and industry observers noted in 1985: "The Army has both boots planted firmly in the electronic age, following the path blazed by the Air Force 15 years ago in fielding weapons less mechanical and more reliant on computers and microcircuits" (Atkinson and Hiatt 1985). Electronics and computer technology were also being diffused throughout other military applications, in simulators for training and in simulations analysis and technology development, widespread in the military by the end of the 1970s (Chait et al. 2007).[19] The

19. In this defense report, "Project Hindsight Revisited," it is noted that "the Abrams program [in the 1970s] called for gun ranges, armor testing ranges, special facilities for testing armor and munitions containing depleted uranium (DU), test tracks, materials laboratories, and visualization techniques for measuring the behavior of munitions at very high speeds and

American Society for Mechanical Engineers reportedly offered seminars for "engineering professionals [to] learn how to get in on the ground floor of the state-of-the-art cornucopia" (Hiatt and Atkinson, 1985). A U.S. investment banking defense analyst said (as cited in DiFilippo [1990, 117]), "SDI is the future of the defense industry. No competitive high-tech company can afford not to be part of SDI."

The shift to employment in the electronics occupations *within* the military (this includes electronics engineers, but also other electronics occupations) continued the post–World War II trend, though with a sharp increase in the late 1970s, with electronics occupations accounting for one in twenty enlisted jobs at the end of World War II to one in five enlisted jobs by the early 1980s, and it "accounts for all the growth in the technical occupations since 1957 . . . the increase in the proportion of electronics technicians in the armed forces closely parallels changes in the electronics content of military equipment (as measured by cost), which has grown from an estimated 10 to 20 percent in the 1950s, to 20 to 30 percent in the late 1960s/early 1970s, and to nearly 40 percent by 1983" (Binkin 1988, 188). By the late 1980s, 45 percent of all engineers working in defense were electrical engineers (as compared to only 28 percent of nondefense engineers who were electrical engineers [U.S. Congress, Office of Technology Assessment 1992, 104]).

Important to note is that the increase in the military was occurring while there were declines in civilian-sector employment; the military increases absorbed declines in other industries, thus increasing the share of military-dependent employment within all engineering employment. Although this countercyclical employment was occurring throughout engineering fields, it was a particularly sharp change in electrical engineering and computer science fields because of the dramatic change in the types of technology being developed, toward greater use of electronics. For example, during the airline industry's financial problems in the 1970s through the 1980s, civilian aerospace purchases were in steep decline and the large increase in military aerospace development and acquisition led to only a 2 percent overall increase in aerospace output during the early 1980s. However, "the defense share of aircraft output equaled 66 percent in 1985, compared with 43 percent in 1977. The aircraft and missile engine industry showed a similar increase in defense market share, rising from 47 to 78 percent between 1977 and 1985" (Henry and Oliver 1987, 6).

The semiconductor industry was a key industry in the electronic component supply chain for these military systems and, in fact, much of the electronic technology development (systems, components, and overall R&D)

---

during penetration of targets. Also, much of the work in the four systems we studied relied on advanced computers for modeling physical phenomena, such as the aeromechanics of the helicopter, finite element analysis of the composite sabot for the Abrams' kinetic energy rounds, and firing tests" (Chait et al. 2007, 20).

was the result of defense department investment during the 1970s. In addition, although the overall economy was stagnant throughout this period, it was a period of rapid growth for the semiconductor and electronics industries and the growth of the West Coast and Southwest electronics firms: "In just two years, from 1978 to 1980, the nation's semiconductor manufacturing capacity doubled" (Saxenian 1996, 87). During the first part of the 1980s, it was the semiconductor and microcomputer industry that drove much of the private-sector growth in the West and Southwest, whereas in the East Coast, particularly in the Route 128 region, the late 1970s and early 1980s was the height of the minicomputer industry. Between 1975 and 1985, Silicon Valley added nearly 150,000 jobs (total employment) before employment leveled off and then declined; in the Route 128 area, about 100,000 jobs were added during this period and then employment declined sharply (Saxenian 1996, 3, 99).

In the mid-1980s, several changes led to a dramatic reversal of fortune for the electronics industry. In semiconductors, foreign firms expanded their market share, with U.S. memory producers market share declining in 1986 to one-third of their dominant 75 percent market share in 1980 (Brown and Linden 2011, 17) and similar declines in other segments of chip manufacturing.[20] Although U.S. firms' actual production still increased through the 1980s, the declining market share and increased productivity (increasing annually by 13 percent from 1986 to 1992) presumably created caution about future domestic expansion and hiring in the face of intensifying foreign competition and their steep decline in market share; as one analyst report explains, "U.S. firms increased output without increasing employment by adopting new technologies that made U.S. semiconductor manufacturing more capital-intensive and productive and by transferring labor-intensive manufacturing abroad" (Malison 1993, 6). In Silicon Valley, the employment decline started in 1985 and occurred rapidly; 20 percent of all semiconductor employees lost their jobs and only three firms remained in dynamic random-access memory production by 1986, leading to "the worst recession in Silicon Valley history" (Saxenian 1996, 89; DiFilippo 1990, 64). From 1984 to 1995 the computer-manufacturing industry lost 32 percent of its workforce, falling at an annual rate of 3 percent (Warnke 1996, 18–23; Moris 1996). The minicomputer industry, based in the Route 128 region outside Boston, went into steep decline as demand shifted from minicomputers to personal computers and workstations. Although Silicon

20. "From a leading share of almost 62 percent in 1980, U.S. chipmakers lost roughly 25 percent of the global market over the next nine years, declining to a low point of 37 percent by 1989. Japanese semiconductor firms by 1989 accounted for more than half of global semiconductor revenues . . . Japanese semiconductor equipment manufacturers increased their global market share from less than 20 percent in 1980 to almost 50 percent in 1990, largely at the expense of U.S. equipment firms, whose market share declined from roughly 75 percent to less than 45 percent during the same period" (VLSI Research 1998, cited in National Research Council [1999], 251–52).

Valley recovered in the late 1980s, elsewhere the loss was permanent in the minicomputer industry and some other segments of the semiconductor and consumer electronics industries.

The decline of the semiconductor industry as market share was lost to foreign competitors, the decline of the minicomputer industry, and the end of R&D funding for Star Wars[21] all happened in the mid-1980s. It was thus the shifting fortunes of several industries occurring around the same time that led to the precipitous drop in demand for electrical engineering and computer science graduates.

Although the decline in engineering and computer science graduates precedes the decline in overall Star Wars spending, as noted above engineering is more dependent on R&D spending patterns than on the overall budget. Research and development spending may diverge from overall spending in preceding both the increases and the declines since engineering will increase before production and then may decline well before decreases in production and overall spending.[22] The demand for new electrical engineering and computer science graduates may have been satisfied during the rapid expansion of the 1982–1984 period with little or no additional hiring after that period. The 1986 peak in electrical engineering graduates would be consistent with the cobweb model of labor market adjustment, lagging by about two years changes in the actual labor market demand. The rise and decline of electrical engineering bachelor's degrees coincides with trends in salaries earned by both electrical engineering and computer science majors, which increased in the early 1980s and fell in the mid-1980s.

The most notable characteristic in the trends of civil, chemical, and other engineering bachelor's degrees during this time period was a modest business cycle procyclicality, particularly in the last thirty years, with peaks in the mid-1980s and late 1990s and troughs in 1990 and 1991. The procyclicality of the supply of graduates in these fields is attributable to at least two factors. First, college enrollment increases during periods of macroeconomic weakness as youth seek alternatives to poor employment prospects. These

21. The majority of the 1980s Star Wars and other military growth in R&D spending and increase in scientific and technical personnel in the military was during the 1982–1984 period, and "the rapid military buildup that began in the early 1980s and that includes increasing expenditures for SDI, had a very noticeable growth effect on scientific and engineering employment . . . defense requirements represent a significant fraction of overall employment in high technology industries" (DiFilippo 1990, 120).

22. For example, "[m]any of the biggest defense programs of the 1980s (e.g., General Dynamics's F-16 and McDonnell Douglas's F-15 fighter aircraft for the Air Force, Grumman's F-14 for the Navy, and General Dynamics's M1A2 tank) are coming to an end and few new programs are on the horizon to replace them, which means engineers can be let go while many production workers are still needed. Also, engineers are more heavily affected by the termination of new systems in their development stage. For example, the cancellation of the Navy's next generation attack jet, the A-12, caused the immediate dismissal of 7,000 workers, half of whom were engineers. In this case, the engineers were laid off before most of the production workers were even assigned to the program" (U.S. Congress, Office of Technology Assessment 1992, 104).

enrollees eventually graduate, usually four to six years later, in a labor market that is stronger as a result of economic recovery. Degree completion trends, as opposed to countercyclical enrollment trends, are therefore relatively pro-cyclical. However, anticipation of future growth can also attract students to fields that provide skills in demand during economic recoveries. Civil and chemical engineering track the overall business cycle quite closely, but in other industry fields occupation-specific cycles diverge from the overall economic cycle and affect engineering labor markets and trends that may not mirror the overall economy.

The relative prominence of various engineering fields has shifted over time. While far more electrical engineers were supplied by four-year degree programs in the 1980s than any other field, in the last several years of available data nearly as many students graduated with a bachelor's degree in civil engineering as in electrical engineering (15,900 and 18,400, respectively) in 2013. Mechanical engineering degrees actually exceeded electrical engineering degree awards by 2008. In 1987, a peak of 36 percent of all engineering bachelor's degrees were in electrical engineering and over three times as many engineering graduates were earning an electrical engineering degree as a civil engineering degree. By 2013, 21 percent of all engineering graduates earned a degree in electrical engineering. In 2008, mechanical, civil, and chemical engineering degrees alone made up almost half of all engineering bachelor's degrees.

Two smaller fields that are not shown separately in figure 1.4, computer and biomedical engineering, are of interest because of their recent and prospective growth.[23] Neither exhibits the cyclicality of mechanical, chemical, or civil engineering for most of the period since 1980. Instead, each of these fields is characterized by steady secular growth, with particularly strong growth after 2000. Computer engineering degrees peaked in 2004, likely reflecting declining enrollments after the bursting of the dot-com bubble of the late 1990s through about 2001. Computer engineering programs primarily offer training in hardware engineering. Software engineers, who were even more exposed to the dot-com bubble and bust, would be more likely to hold computer science degrees (though only one-quarter of the IT workforce hold a computer science degree, and over one-third do not hold any four-year degree). Software engineers are not considered in this chapter. Unlike computer engineering awards, biomedical awards continued to increase after the dot-com bust and through the 2001 recession, buoyed by advances in genetics such as successful sequencing of the human genome in 2000, strong growth in health care and pharmaceuticals, and increased funding of medical research by the National Institutes of Health. Biomedical engineering degrees have also emerged as an alternate route to medical

---

23. This discussion relies on more detailed IPEDS data, not shown here, that is only available from the late 1980s to the present.

school, due to the development of advanced medical equipment (MRI, CAT, and other diagnostic equipment) that became widely used during this period. Linsenmeier (2003) points out that biomedical engineering majors have one of the highest acceptance rates to medical school of any major.

1.3.2   Graduate Degree Trends

While shifts in demand, cobweb dynamics, the business cycle, and new markets opened by technological advances shaped the fluctuations in bachelor's degrees over the last thirty years, there was much steadier growth in graduate education in engineering. This trend is comparable to trends in master's degrees outside of engineering, which have also grown continuously over the last fifty years.[24] Since 1966, the number of master's degrees in engineering awarded annually has more than tripled, while doctorates have increased by more than 75 percent. Master's degrees increased steadily since the late 1970s, with the exception of a drop following the post–Cold War decline in U.S. military spending and steep rise and subsequent decline lagging the dot-com bubble. Trends in master's awards in the same engineering field highlighted for bachelor's degrees are presented in figure 1.6. The difference between the trajectories of master's degrees and bachelor's degrees in electrical engineering is perhaps the starkest, with a reasonably steady increase in master's degrees in the field during periods of considerable declines in bachelor's degrees. It is likely that the increase in master's degrees reflected the poor job market and bachelor's degree graduates continuing their education because they were unable to find jobs. There is a slight cyclicality in the mechanical and civil engineering award trends, but a pattern of steady growth is found consistently across all fields.

A graduate degree in engineering is different from a graduate degree in other sciences in that it functions more as a professional, and less as an academic, degree than is the case in nonengineering fields. Individuals holding graduate degrees in engineering are much less likely to work in academia than those holding graduate degrees in biology, chemistry, and physics. Twenty-seven percent of those with doctorates in engineering in 2010 were employed in postsecondary education, compared to 47 percent of those with a doctorate in one of the natural sciences. The disparity is comparable for master's degrees. Among the population of terminal engineering master's degree holders, almost 7 percent worked in postsecondary education in 2008 (with 82 percent in the private sector), while almost 21 percent of those holding terminal master's degrees in the natural sciences work in postsecondary education. Terminal master's degree holders working in postsecondary education do research and teach at high rates, with holders of master's degrees in engineering somewhat more likely to be doing research and less likely

24. Master's degrees generally more than doubled between 1980 and 2008, from 299,095 to 631,711 awards.

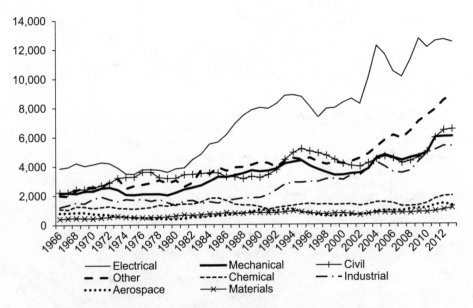

Fig. 1.6   Engineering master's degree awards, 1966–2013
Source: Authors' calculations from engineering degrees in the IPEDS, 1966–2013.

to be teaching than their natural science counterparts. It is unclear from the data, although quite plausible, that master's degree holders working in postsecondary education are primarily employed by community colleges.[25]

1.3.3 Engineering Education Demographics

Engineering is often noted as a field that has made slow progress in achieving racial, ethnic, and gender diversity. Some STEM fields such as science and math have made significant gains in gender parity to the extent that most science fields are at or above gender parity and math is close to parity, with women consistently obtaining between 42 and 48 percent of bachelor's math degrees since the 1970s. By contrast, the share of historically underrepresented African Americans, Hispanics, and other minorities has increased more slowly. The racial and ethnic composition of engineering bachelor's awards is provided in figure 1.7, with the white non-Hispanic share measured on the right axis, and all other categories on the left axis. Whites made up the majority of engineering graduates, although their share declined steadily from over 75 percent in 1989 to under 65 percent in 2013. At the bachelor's degree level, international students comprise a steady 6 to 8 percent share of engineer degrees awarded. The decline in the white non-Hispanic share was counterbalanced by growth in domestic Asian and Hispanic students, and

25. All data are from authors' calculations from the 2008 SESTAT.

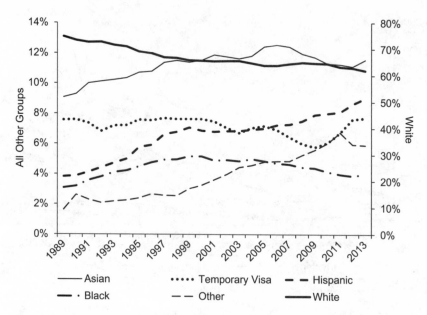

Fig. 1.7   Race and ethnicity composition of engineering bachelor's degree awards, 1989–2013

Source: Authors' calculations from engineering degrees in the IPEDS, 1989–2013.

students of other races or ethnicities. In 2013, African American students made up the smallest share of engineering graduates out of all the groups represented in figure 1.6, comprising between 4 and 5 percent over the past two decades. The largest nonwhite share of bachelor's degrees are held by Asians, at 9 to 12 percent, with small but steady increases by Hispanics, rising from 4 to 9 percent since 1989.

Although engineering has the lowest percentage of women of all STEM fields, with less than one-fifth of all engineering bachelor's degrees going to women, there is quite a large variation by engineering field. Table 1.8 provides the female share of bachelor's awards in the same selected subfields graphed in figures 1.2 and 1.4, as well as three other subfields of particular interest due to their very high or very low female shares. The two largest fields, electrical and mechanical engineering, both have relatively low female shares of just under 12 percent. In civil engineering and "other" engineering categories, women obtain one-fifth to one-quarter of degrees awarded. However, the share of "other" engineering awards going to women obscures broad variation across subfields in this category. Computer engineering, for example, has very low female representation at under 10 percent. In contrast, biomedical and environmental engineering have the highest female representation of all engineering, at 39 and 45 percent, respectively.

The share of foreign students (those on F-1 student visas) in engineering

Table 1.8   **Female share of engineering fields**

Engineering field	Female share of bachelor's awards (%)
Electrical	11.76
Mechanical	11.99
Civil	21.44
Other	24.11
Biomedical	38.93
Computer	9.91
Environmental	45.05
Chemical	32.27
Industrial	29.77
Aerospace	13.72
Materials	29.76
Total	19.36

Source: Authors' calculations from the 2013 IPEDS.

varies dramatically between undergraduate and graduate degree awards. At the undergraduate level, foreign students have comprised just under 10 percent of all engineering bachelor's degrees for over two decades. At the master's and doctoral levels, however, the share has increased from under 30 percent to over 40 percent at the master's level, and from just under 50 percent to around 60 percent in recent years at the doctoral level. The number of degrees at each level varies (see figure 1.8); in 2013, foreign students received about 7,000 bachelor's degrees, just under 18,000 master's degrees, and about 2,400 PhDs. The large number of master's degrees awarded to foreign students reflects, in part, the migration of students who receive a bachelor's degree in their home country and then enter a U.S. master's degree program both to obtain an engineering degree that may better qualify them for employment and also as the "entry portal" into the U.S. labor market. For most U.S. students, the bachelor's degree is the terminal degree and sufficient for entry into the engineering labor market.

1.4   Conclusion

Engineers are fundamental to a well-functioning, developed economy. The generation and maintenance of modern technology and the infrastructure that supports the nation would be unthinkable without engineers, who operate as the intermediary between scientific advances and improvements in everyday life. As such, researchers and policymakers are justifiably interested in the functioning of the labor market for engineers, and particularly whether the transition from school to work for new engineers is operating smoothly and providing a reliable supply to industry and government.

This chapter provides an overview of these issues. The first section con-

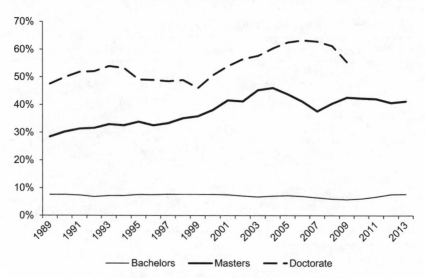

Fig. 1.8   Share of foreign engineering graduate student (F-1) visas, 1989–2013
Source: Authors' calculations from engineering degrees in the IPEDS, 1989–2013.

sidered some major factors in the demand for engineers, which varies widely
by field and industry. Some fields, such as civil and electrical engineering, are
in demand across a wide range of industries, while others such as chemical
engineering are concentrated in a single industry, in many cases the manufacturing sector. This differential distribution of engineers across industries has
important consequences for the response of the engineering labor market
to vicissitudes of the business cycle. Over the last thirty years an increasing
share of engineers have been employed in independent engineering firms,
reflecting broader trends of outsourcing in the economy.

While most engineers are employed in the private sector, government is an
important force in the labor market for certain types of engineers. Many civil
and environmental engineers, for example, are employed in infrastructure
projects that are publicly funded. Alternatively, specialized engineering fields
like aerospace and biomedical engineering are often heavily dependent on
federal grants and contracts, if not the beneficiaries of direct federal employment. Finally, the replacement of aging workforces does not appear to be a
critical factor in the overall demand for engineers. Aging workforces are generally workforces that are in decline, and employment growth for younger
workers tends to be in growing industries rather than for replacement. While
replacement of retiring workers is always occurring, aging workforces are
typically not a sign of impending replacement demand as the source of
increasing demand for new graduates.

The next section considered the production of new engineers by colleges and universities. The pattern of engineering bachelor's degree awards

varied substantially across fields. Civil and mechanical degrees, for example, exhibited a strong cyclicality that reflected the business cycle, electrical engineering showed much more dramatic shifts over time—most notably, a steep decline in the production of electrical engineering students in the 1980s—that were closely correlated with declining labor market opportunities for electrical engineers but did not reflect the overall business cycle. A few smaller fields, like computer and biomedical engineering, showed consistent secular growth at the undergraduate level. In contrast with bachelor's degrees, graduate degree awards showed persistent gains over the last thirty years.

In the midst of often heated policy debates about engineering labor shortages it is useful to take a step back and explore the functioning of the engineering labor market piece by piece: the demand for engineers, the supply of new engineers, and the institutional environment that produces new engineers. Not surprisingly each of these pieces of the labor market has exhibited both consistency and change over the last thirty years, and a steady eye on each of them is required for a clear understanding of the experiences of American engineers going forward.

References

Allen, R. G. D., and Brinley Thomas. 1939. "The Supply of Engineering Labour under Boom Conditions." *Economic Journal* 49 (194): 259–75.

Arrow, Kenneth J., and William M. Capron. 1959. "Dynamic Shortages and Price Rises: The Engineer-Scientist Case." *Quarterly Journal of Economics* 73 (2): 292–308.

Atkinson, Richard C., and Fred Hiatt. 1985. "Military in a Fix." *Washington Post*, Aug. 18.

Biddle, Jeff, and Karen Roberts. 1994. "Private Sector Scientists and Engineers and the Transition to Management." *Journal of Human Resources* 29 (1): 82–107.

Binkin, Martin. 1988. "Technology and Skills: Lessons from the Military." In *The Impact of Technological Change on Employment and Economic Growth, Papers commissioned by the Panel on Technology and Employment of the National Academy of Sciences, National Academy of Engineering, and Institute of Medicine*, edited by Richard M. Cyert and David C. Mowery, 185–222. Cambridge, MA: Ballinger Publishing Company.

Blank, David M., and George J. Stigler. 1957. *The Demand and Supply of Scientific Personnel*. New York: National Bureau of Economic Research.

Board of Governors of the Federal Reserve System. 2011. *Industry Production and Capacity Utilization—G.17*. Table 4: Revised Data for Capacity and Utilization for Total Industry, 1980–2010. Washington, D.C.: Federal Reserve System.

Brown, Clair, and Greg Linden. 2011. *Chips and Change: How Crisis Reshapes the Semiconductor Industry*. Cambridge, MA: MIT Press.

Bureau of Labor Statistics. 2015. "Usual Weekly Earnings of Wage and Salary Workers Fourth Quarter 2014." News Release (USDL-15-0071). https://www.bls.gov/news.release/archives/wkyeng_01212015.pdf.

Bush, Vannevar. 1945. *Science, the Endless Frontier: A Report to the President on a Program for Postwar Scientific Research.* Washington, D.C.: National Science Foundation.

Chait, Richard, John Lyons, Duncan Long, and Albert Sciarretta. 2007. *Lessons from Project Hindsight Revisited.* Washington, D.C.: Center for Technology and National Security Policy at the National Defense University.

Congressional Budget Office. 2008. "Factors underlying the Decline in Manufacturing Employment Since 2000." *Economic and Budget Issue Brief*, CBO, Washington, D.C.

DiFilippo, Anthony. 1990. *From Industry to Arms: The Political Economy of High Technology.* New York: Greenwood Press.

Fox, J. Ronald. 2011. *Defense Acquisition Reform, 1960–2009: An Elusive Goal.* Washington, D.C.: Center of Military History, U.S. Army.

Freeman, Richard. 1975. "Supply and Salary Adjustments to the Changing Science Manpower Market: Physics, 1948–1973." *American Economic Review* 65 (1): 27–39.

———. 1976. "A Cobweb Model of the Supply and Starting Salary of New Engineers." *Industrial and Labor Relations Review* 29 (2): 236–48.

———. 2007. "Is a Great Labor Shortage Coming? Replacement Demand in the Global Economy." In *Reshaping the American Workforce in a Changing Economy*, edited by Harry Holzer and Demetra Nightingale, 3–24. Washington, D.C.: The Urban Institute Press.

Gibson, G., Jr., A. Davis-Blake, K. Dickson, and B. Mentel. 2003. "Workforce Demographics among Project Engineering Professionals—Crisis Ahead?" *Journal of Management in Engineering* 19 (4): 173–82.

Hansen, W. Lee. 1967. "The Economics of Science and Engineering Manpower." *Journal of Human Resources* II (2): 191–220.

Henry, David, and R. Oliver. 1987. "The Defense Buildup, 1977–85: Effects on Production and Employment." *Monthly Labor Review* 110:3–11.

Hiatt, Fred, and Richard C. Atkinson. 1985. "Strategic Defense, the Seeds of a New Industry are Planted." *Washington Post*, Oct. 20.

Kaiser, David. 2002. "Cold War Requisition, Scientific Manpower, and the Production of American Physicists after World War II." *Historical Studies in the Physical and Biological Sciences* 33:131–59.

Linsenmeier, R. A. 2003. "What Makes a Biomedical Engineer?" *IEEE Engineering in Medicine and Biology Magazine* 22 (4): 32–38.

Lowell, B. Lindsay, and Hal Salzman. 2007. *Into the Eye of the Storm: Assessing the Evidence on Science and Engineering Education, Quality, and Workforce Demand.* Washington, D.C.: The Urban Institute.

Lynn, Leonard, and Hal Salzman. 2010. "The Globalization of Technology Development: Implications for U.S. Skills Policy." In *Transforming the U.S. Workforce Development System: Lessons from Research and Practice*, edited by David Finegold, Mary Gatta, Hal Salzman, and Susan Schurman. Ithaca, NY: ILR Press.

Malison, Andrew F. 1993. *Industry & Trade Summary: Semiconductors.* Washington, D.C.: U.S. International Trade Commission, Office of Industries.

McPherson, Michael S. 1986. "Modeling the Supply of Scientists and Engineers: An Assessment of the Dauffenbach-Fiorto Work." In "The Impact of Defense Spending on Non-Defense Engineering Labor Markets: A Report to the National Academy of Engineering," 43–51. Washington, D.C., National Academy Press.

Moris, Francisco. 1996. "Semiconductors: The Building Blocks of the Information Revolution." *Monthly Labor Review* 119 (8): 6–17.

National Academy of Sciences, National Academy of Engineering, and Institute of Medicine. 2007. *Rising above the Gathering Storm: Energizing and Employing America for a Brighter Economic Future*. Washington, D.C.: National Academies Press.

National Research Council. 1986. "The Impact of Defense Spending on Non-Defense Engineering Labor Markets: A Report to the National Academy of Engineering." Washington, D.C., National Academy Press.

———. 1999. "U.S. Industry in 2000: Studies in Competitive Performance." Washington, D.C., National Academy Press.

———. 2012. *Assuring DoD a Strong Science, Technology, Engineering, and Mathematics (STEM) Workforce*. Washington, D.C.: National Academies Press.

National Science Board. 2014. *Science and Engineering Indicators 2014*. Arlington, VA: National Science Foundation (NSB 14-01).

National Science Foundation. 1953. *Third Annual Report of the National Science Foundation*. Washington, D.C.

———. 1995. *Federal Scientists and Engineers: 1989–93*, NSF 95-336, Arlington, VA.

———. 2012. *Science and Engineering Indicators 2012*. Figures 3-10 and 3-11. Arlington, VA. http://www.nsf.gov/statistics/seind12/.

National Science Foundation, Division of Science Resources Statistics. 2005. *Federal Scientists and Engineers: 1998–2002*, NSF 05-304, Project Officer John Tsapogas, Arlington, VA.

———. 2008. *Federal Scientists and Engineers: 2003–05*. Detailed Statistical Tables NSF 09-302, Arlington, VA. http://www.nsf.gov/statistics/nsf09302/.

National Science Foundation, National Center for Science and Engineering Statistics. 2011. *Science and Engineering Degrees: 1966–2008*. Detailed Statistical Tables NSF 11-316, Arlington, VA. http://www.nsf.gov/statistics/nsf11316/.

President's Council on Jobs and Competitiveness. 2011. "Taking Action, Building Confidence." Interim Report, October. http://files.jobs-council.com/jobscouncil/files/2011/10/JobsCouncil_InterimReport_Oct11.pdf.

President's Science and Advisory Committee. 1962. *Meeting Manpower Needs in Science and Technology. Report no. 1: Graduate Training in Engineering, Mathematics, and Physical Sciences*. Washington, D.C.: U.S. Government Printing Office.

Ryoo, Jaewoo, and Sherwin Rosen. 2004. "The Engineering Labor Market." *Journal of Political Economy* 12 (1): S110–40.

Salzman, Hal. 2013. "What Shortages? The Real Evidence about the STEM Workforce." *Issues in Science and Technology* 28 (4): 58–67.

———. 2015. "Testimony on Science and Engineering Labor Demand and Supply and High-Skill Guestworker Programs." Submitted to the Senate Committee on the Judiciary, Washington, D.C., Mar. 17. http://dx.doi.org/doi:10.7282/T3ZK5JC3.

Salzman, Hal, Daniel Kuehn, and B. Lindsay Lowell. 2013. "Current and Proposed High-Skilled Guestworker Policies Discourage STEM Students and Grads from Entering IT." *Economic Snapshot*, May 30.

Saxenian, AnnaLee. 1996. *Regional Advantage: Culture and Competition in Silicon Valley and Route 128*. Cambridge, MA: Harvard University Press.

Teitelbaum, Michael. 2014. *Falling Behind? Boom, Bust, and the Global Race for Scientific Talent*. Princeton, NJ: Princeton University Press.

U.S. Census Bureau, American Community Survey. 2012. *Employment in STEM Occupations by Field of Degree for the First Listed Bachelor's Degree: 2012*. https://census.gov/people/io/files/Table%205.%20STEM.xlsx.

U.S. Congress, Office of Technology Assessment. 1992. *After the Cold War: Living with Lower Defense Spending*, OTA-ITE-524. Washington, D.C.: U.S. Government Printing Office.

Warnke, Jacqueline. 1996. "Computer Manufacturing: Change and Competition." *Monthly Labor Review* 119 (8): 18–30.
Yuskavage, Robert E., Erich H. Strassner, and Gabriel W. Medeiros. 2009. "Outsourcing and Imported Services in BEA's Industry Accounts." In *International Trade in Services and Intangible in the Era of Globalization*, edited by Marshall Reinsdorf and Matthew J. Slaughter. Chicago: University of Chicago Press.

I

Engineering Education

2

Career Plans of Undergraduate Engineering Students
Characteristics and Contexts

Shannon K. Gilmartin, anthony lising antonio,
Samantha R. Brunhaver, Helen L. Chen, and
Sheri D. Sheppard

Each year some 80,000 students graduate from one of over 350 universities and colleges in the United States with a bachelor of science (BS) degree in engineering (American Society for Engineering Education [ASEE] 2015; National Science Foundation [NSF] 2015). This chapter examines how many of these graduates see this degree as leading to engineering work, the types of engineering students who are more likely to go into engineering work, and the experiences that influence their entering the profession or working elsewhere. The chapter builds on findings from the Academic Pathways of People Learning Engineering Survey (APPLES) (Sheppard et al. 2010),[1]

Shannon K. Gilmartin is senior research scholar at the Michelle R. Clayman Institute for Gender Research and adjunct professor in mechanical engineering at Stanford University. anthony lising antonio is associate professor of education at Stanford University and associate director of the Stanford Institute for Higher Education Research. Samantha R. Brunhaver is assistant professor of engineering at The Polytechnic School at Arizona State University. Helen L. Chen is senior researcher in the Designing Education Lab in the Center for Design Research within the Department of Mechanical Engineering at Stanford University. Sheri D. Sheppard is the Burton J. and Deedee McMurtry University Fellow in Undergraduate Education and professor of mechanical engineering at Stanford University.

Research supported by grants from the National Science Foundation (grant nos. ESI-0227558 and ESI-1022644) and the Alfred P. Sloan Foundation, and the Stanford Graduate Fellowship program. Any opinions, findings, conclusions, or recommendations expressed in this publication are those of the authors and do not necessarily reflect those of NSF, Sloan, Stanford, or NBER. The authors wish to thank the editors for their helpful comments on earlier versions of this chapter. For acknowledgments, sources of research support, and disclosure of the authors' material financial relationships, if any, please see http://www.nber.org/chapters/c12686.ack.

1. Prior work identified positive predictors of plans to pursue engineering jobs after college, including financial motivation to study engineering (i.e., seeing engineering as a means to a job, reporting that "engineers are well paid"), psychological motivation to study engineering (i.e., being intrinsically interested in engineering, reporting that engineering is "fun"), exposure to engineering through internships, visits, and employment, and involvement in engineering

which was a part of the NSF-funded Academic Pathways Study (APS).[2] We augment the APPLES data with data from three national sources. We employ multilevel modeling techniques to consider the effects of individual-level and institution-level measures simultaneously.

Our study fits into a larger body of work on engineering career pathways, which has studied the factors influencing persistence and retention in undergraduate engineering programs[3] and the intentions of seniors to take engineering jobs upon graduation.[4] These studies point to a link between students' perceptions of educational environments in engineering and of their own engineering skills and preparedness, and their plans to continue on in an engineering career.[5]

The current work builds on this literature of how engineering students conceive of their professional futures to examine the variation of student plans with individual and environmental characteristics such as socioeconomic background, major field of study, institutional selectivity, and

courses. Negative predictors included confidence in professional and interpersonal skills and extracurricular involvement in nonengineering activities. Across all of our models of postgraduation plans, confidence in professional/interpersonal skills consistently characterized students who were leaning away from engineering options and toward nonengineering options, while psychological motivation to study engineering consistently characterized students who were leaning toward engineering options and away from nonengineering options.

2. The Center for the Advancement of Engineering Education (CAEE) was established by NSF in 2003. CAEE's five partner institutions were Colorado School of Mines, Howard University, Stanford University, University of Minnesota, and University of Washington (the lead institution). CAEE consisted of three research components: Scholarship on Learning Engineering, Scholarship on Teaching Engineering, and the Institute for Scholarship on Engineering Education. The APS was a major undertaking of the Scholarship on Learning Engineering component (Clark et al. 2008).

3. Seymour and Hewitt (1997), Brainard and Carlin (1998), Eris et al. (2005, 2007, 2010), Huang, Taddese, and Walter (2000), Besterfield-Sacre, Atman, and Shuman (1997), and Besterfield-Sacre et al. (2001). Adelman (1998) used students' college transcripts in the High School and Beyond/Sophomore Cohort Longitudinal Study (1982–1993). Ohland et al. (2008, 2009, 2011) and Lord et al. (2008, 2009) analyzed the Multiple-Institution Database for Investigating Engineering Longitudinal Development (MIDFIELD) data set of academic records for over 75,000 first-time-in-college students matriculating in engineering from 1988 through 1998 across nine partner institutions.

4. Margolis and Kotys-Schwartz (2009) found that 65 percent of seniors at a midwestern university had some reservations about committing to an engineering career after college. In a study of 1,629 freshmen through seniors across nine institutions, Amelink and Creamer (2010) found a relationship between satisfaction with instruction and plans to work in engineering. Investigating decisions to pursue six different engineering and nonengineering career options within three years of graduation, Ro (2011) found that students who were more likely to pursue engineering work or graduate school options tended to be male, to report greater curricular emphases on core engineering thinking and professional skills in their engineering programs, to have had more active and collaborative learning experiences, and to rate their engineering and design skills highly. By contrast, students intending to work outside of engineering were more likely to report greater curricular emphasis on professional values and to have higher confidence in their understanding of the relationship between engineering and social, economic, and other contexts. Seniors compared with sophomores and students enrolled in general engineering majors compared with mechanical engineering were more likely to be considering nonengineering careers.

5. Sheppard et al. (2014) give a deeper methodological review of research on plans and pathways.

labor market conditions. It also illuminates engineering career paths beyond graduation. As Sheppard et al. (2014) note, most studies report the proportions of graduates who remain in engineering professions (e.g., Bradburn et al. 2006; Choy, Bradburn, and Carroll 2008; Forrest Cataldi et al. 2011; Regets 2006), by gender (e.g., Frehill 2007a, 2007b; Robst 2007), or within major (Reese 2003). In analyses of persistence across all science, math, technology, and engineering fields by students' achievement level measured by SAT math scores or college GPA, depending on the time point under study, Lowell et al. (2009) observed that in recent decades, the highest-performing students and graduates were leaving science and engineering pathways at a greater rate than were lower-performing students and graduates. The data for the current study highlight additional pre-labor-force factors that may be important in differentiating professional pathways.

2.1 Data

The data for this study are drawn primarily from APPLES, a fifty-item survey instrument administered to over 4,000 students across twenty-one U.S. colleges and universities in 2008 (Chen et al. 2008; Donaldson et al. 2007, 2008). Clark et al. (2008) detail the design of the larger APS and related instrumentation. Institutions were recruited to participate in the APPLES study based on a stratified sampling plan designed to capture the broad range of four-year colleges and universities in the United States (Donaldson et al. 2008). Appendix A shows the distribution of institutions by 2000 Carnegie Classification category (Carnegie Foundation for the Advancement of Teaching 2001). Institutions were geographically diverse, representing seventeen states across major U.S. regions. Sheppard et al. (2010) provide further discussion of the representativeness of these twenty-one institutions. The final Center for the Advancement of Engineering Education (CAEE) report (Atman et al. 2010) lays out major project findings.

We sample 2,143 survey respondents identifying as "juniors," "seniors," or "fifth-year seniors or more." Respondents in each academic cohort were similar on key demographic measures and measures of engineering plans following graduation, thus allowing us to aggregate cohorts and consider predictors of plans across a larger sample. Women make up 30.8 percent of the sample and thus are overrepresented compared with the U.S. population of students who earned an engineering degree in 2008. Additionally, (prospective) electrical engineering degree earners are slightly underrepresented and mechanical engineering degree earners are slightly overrepresented.[6]

We merge three additional sources of data into the APPLES survey data

6. All APPLES respondents were enrolled at their undergraduate institutions at the time of the study; thus, "degree earners" is a prospective label. See NSF (2011a) for statistics on the 2008 population of engineering degree earners by gender and major. Sheppard et al. (2010) examine the representativeness of the student respondent sample relative to the national population of engineering undergraduates.

set to analyze institutional and labor market effects: median annual wage data by engineering field and U.S. state from the Bureau of Labor Statistics' (BLS) National Cross-Industry Estimates, May 2007 (BLS 2011); and data on institutional selectivity and resources for each of the twenty-one APPLES institutions from the Integrated Postsecondary Education Data System (IPEDS), 2006 (National Center for Education Statistics [NCES] 2011), and the ASEE Engineering and Engineering Technology College Profiles, 2006 (ASEE 2011). For measures of selectivity (i.e., SAT percentile scores) at a small number of schools, 2007–2008 IPEDS data were used in the absence of such data for 2006–2007; for the student-to-faculty ratio at each institution, 2008–2009 IPEDS data were used in the absence of such data for 2006–2007.

2.1.1 Hierarchical Linear Modeling

Due to the nested students-within-institutions nature of the data set, hierarchical linear modeling (HLM) techniques were used for multivariate modeling. Ordinary least squares regression techniques may underestimate standard errors in this type of sample, given the dependence of observations within each school. Hierarchical linear modeling accounts for such clustering by partitioning between-school variance and within-school variance, thereby generating more robust standard errors and allowing us to directly estimate school-level effects on outcomes (Raudenbush and Bryk 2002). The outcome measure of interest is respondents' postgraduation plans, which we define by three mutually exclusive categories: engineering-focused plans, nonengineering-focused plans, and mixed plans. We conducted multinomial logistic regression (MLR) analyses with a multilevel specification (hierarchical generalized linear modeling) of this outcome measure.

Level 1 models focus on the relationships between students' postgraduation plans, background characteristics, educational experiences and attitudes, and education and employment contexts. The models contain eighteen independent variables.[7] Each independent variable was centered around its group mean. To facilitate interpretation and comparability of coefficients, we standardized all nondichotomous variables prior to centering. Prelimi-

---

7. The multinomial logistic model in a hierarchical generalized linear model is given by

$$\text{Prob}(R_{ij} = 1) = \phi_{1ij}$$
$$\text{Prob}(R_{ij} = 2) = \phi_{2ij}$$
$$\text{Prob}(R_{ij} = 3) = \phi_{3ij} = 1 - \phi_{1ij} - \phi_{2ij}$$

where each response category I is associated with a probability (ϕ). Using a multinomial logit link (logged ratio of probabilities), the Level 1 structural model is expressed as

$\eta_{1ij} = \beta_{0j(1)} +$ [student background characteristics (2 measures)
   + educational experiences/attitudes (7 measures) + major (8 measures)
   + expected salary (1 measure) in the form $\Sigma\beta_{qj(1)}X_{qij}$]

$\eta_{2ij} = \beta_{0j(2)} +$ [student background characteristics (2 measures)
   + educational experiences/attitudes (7 measures) + major (8 measures)
   + expected salary (1 measure) in the form $\Sigma\beta_{qj(2)}X_{qij}$]

nary Level 1 models tested the effects of "gender"[8] and "underrepresented racial/ethnic minority (URM) status" as separate, independent measures. Final models tested "URM women" versus other students based on exploratory descriptive and multivariate results that indicated this group may have unique postgraduation plans (see Sheppard et al. 2010).

Level 2 models, which focus on the effect of differences in institutional selectivity and resources on postgraduation plans, are more difficult to estimate given that there are only twenty-one institutions in the APPLES data set. Preliminary analyses of simple correlations showed how these institutional characteristics clustered into thematic groups and were related to postgraduation plans. These groups formed the basis of the subsequent testing sequence designed to account for limited variation in institutions. Some thematic groups included only one variable, for example, percentage of undergraduates enrolled as part-time students. Other groups comprised two variables (representing academic selectivity, financial aid receipt, enrollment characteristics, and student-faculty ratio), which we first modeled as the relationship between a variable and the outcome alone and then in conjunction with its companion measure (this allowed us to assess if a variable's predictive power was independent of a related measure—and which of the two was the stronger predictor). Ten sets of Level 2 models were constructed according to this heuristic. Each Level 2 variable was centered around the grand mean. As with the Level 1 models, nondichotomous variables were standardized prior to centering.

Finally, post hoc univariate analyses were conducted to elaborate on findings from multivariate models. These included one-way ANOVAs with post hoc multiple comparisons, and Welch robust tests of equality of means where variance was heterogeneous. For pairwise comparisons, we used Bonferroni tests in the case of homogeneous variance, and Games-Howell tests in the case of heterogeneous variance.

2.1.2   Outcome Measures

To measure students' postgraduation plans, the APPLES survey asked students "How likely is it that you would do each of the following after

---

where η is the log-odds of being in category (1) or (2) relative to category (3) as a function of the group mean (β_0) and covariates (X).

At Level 2, the model is given by

$$\beta_{0j(1)} = \gamma_{0(1)} + \Sigma\gamma_{0s(1)}W_{sj} + \mu_{0j(1)}$$
$$\beta_{0j(2)} = \gamma_{0(2)} + \Sigma\gamma_{0s(2)}W_{sj} + \mu_{0j(2)}$$
$$\beta_{qj(m)} = \gamma_{q0(m)}, \; q = 1, \ldots, 17$$

where the group mean is a function of the grand mean (γ_0) and covariates (W).

8. To retain the maximum number of cases for our multivariate models, gender was imputed for twenty missing values using logistic regression techniques, and mean-replacement was used for missing values on other independent variables, including four composite measures (the proportion of missing values on any given variable was low, ranging from < .5 percent to 2.5 percent).

graduation?" and listed four possible options: "work in an engineering job," "work in a nonengineering job," "go to graduate school in an engineering discipline," and "go to graduate school outside of engineering."[9] Each item was measured on a five-point scale, from 0 = "definitely not" to 4 = "definitely yes." Respondents were then placed into one of three analytic groups based on combinations of responses across all four postgraduation categories:

"engineering-focused students" are those who marked probably/definitely yes to one or both engineering options (job, graduate school) and probably/definitely no to both nonengineering options (job, graduate school);
"nonengineering-focused students" are those who marked probably/definitely yes to one or both nonengineering options and probably/definitely no to both engineering options; and
"all other plans" include all other possible response combinations to the four postgraduation survey items.

We developed a second classification to analyze a narrower subset of students falling into the "all other plans" group. In this classification, we replaced "all other plans" by "cross-field plans," which includes only those response combinations where students marked probably/definitely yes to at least one engineering option and one nonengineering option. This "refined sample" reduced the observations for analysis from 2,143 students to 1,318 students.

Level 1 and Level 2 MLR analyses were conducted on the groups defined by each classification scheme. To support interpretation of the multinomial results, each of the four constituent pathways questions (likelihood to "work in an engineering job," "work in a nonengineering job," "go to graduate school in an engineering discipline," "go to graduate school outside of engineering") also was modeled separately with the same Level 1 specification using HLM. We do not present the results of these supplementary analyses in this chapter, but reference them when necessary to clarify discussion of the MLR results. Detailed results of these analyses are available from the authors upon request.

Taking Level 1 measures first, we measured students' self-reported socioeconomic background (SES) by a question: "Would you describe your family as: . . . ?" This item was measured on a five-point scale, from 0 = "low income" to 4 = "high income." We identified nine major fields of study: aerospace engineering, chemical engineering, civil and environmental engineering, computer science/engineering, electrical engineering, industrial engineering, mechanical engineering, "bio-x" engineering, and "other engineering." The majority of institutions in the sample include

9. Students were not presented with a list of specific engineering jobs on the APPLES instrument. Rather, they were asked about "engineering jobs" in the aggregate; the intention was to elicit students' interest in these jobs as they conceived them to be. Interpretive limitations to these measures are discussed in Sheppard et al. (2014).

civil and environmental engineering in one department, so these majors were aggregated into a single major with 280 respondents marking "civil" and twenty-two respondents marking "environmental." Computer science/engineering includes only students reporting "computer science/engineering (in engineering)" as distinct from a computer science program outside of engineering. In addition to the majors listed on the survey, we developed two categories from students' write-in responses to an "other engineering major" open-ended field: "bio-x," which covers bioengineering programs such as bioengineering, biomedical engineering, and so forth; and "other engineering," which includes agricultural engineering, construction engineering, engineering math and physics, engineering operations research and business, general engineering, materials and metallurgical engineering, nuclear engineering, ocean engineering, and other engineering. Two APPLES institutions offer a "general engineering" degree only; all respondents at these two schools are classified as "other engineering" majors. Mechanical engineering is the reference group for all regression models.

Drawing from BLS 2007 median annual earnings data by state, students were assigned an expected earnings based on their engineering major and the state in which their college was located. The BLS engineering fields generally correspond with APPLES major groups, but where possible we assigned students in more detailed majors earnings from more detailed BLS occupations as described in the footnote.[10] We also used several variables that earlier work on the APPLES data set found to be strong predictors of students' postgraduation plans as covariate controls for the models:

Financial motivation. A three-item composite measure of students' financially motivated reasons to pursue engineering study.

Intrinsic psychological motivation. A three-item composite measure of students' "intrinsic" reasons to pursue engineering study.

Exposure to engineering work. Obtained from the survey question "How much exposure have you had to a professional engineering environment as a visitor, intern, or employee?" based on a four-point scale from 0 = "no exposure" to 3 = "extensive exposure."

Involvement in engineering classes.

Participation in nonengineering activities. Obtained from the survey question "How often are you involved in the kinds of nonengineering activities

10. Civil engineering majors were assigned median earnings for civil engineers. Environmental engineering majors were assigned earnings for environmental engineers. Students in computer science/engineering majors were assigned an average median earnings across three BLS fields: computer hardware engineers, computer software engineers: applications, and computer software engineers: systems software. Students in bio-x engineering majors were assigned a median earnings for biomedical engineers. For students in the "other engineering" major group, a "match" to BLS data was made wherever possible, for example, materials engineers in our "other engineering" major group were assigned median earnings for materials engineers in the BLS data. However, median earnings for these "other" fields were not available for every field in every state; in the case of a nonmatch, these students were assigned a median for the BLS "other engineering" category.

described above (hobbies, civic or church organizations, campus publications, student government, social fraternity or sorority, sports, etc.)?"
and measured on a four-point scale from 0 = "never" to 3 = "frequently."
GPA index. A single-item variable reflecting self-reported cumulative grade
point average measured on an eight-point scale from 0 = "C− or lower"
to 7 = "A or A+," which we converted to a 100-point scale.
Professional/interpersonal confidence. A six-item composite measure of
students' self-concept in professional and interpersonal domains.

Responses to survey items comprising each composite measure were
summed, normalized, and multiplied by 100 for reporting purposes prior
to standardizing. See appendix B for a list of constituent items in each composite measure and corresponding Cronbach's alpha values.

As noted earlier, gender and URM status were tested as separate measures in a preliminary series of models; given results from this analysis, we
included a measure for "URM women" in the final analysis as the group
of students whose postgraduation plans diverged most from others. Students classified as URM are those marking American Indian/Alaska Native,
Black/African American, Hispanic/Latino(a), and/or Native Hawaiian/
Pacific Islander racial/ethnic backgrounds.

Level 2 measures that reflected institutional resources and selectivity were
selected from IPEDS and ASEE data sets. These included SAT composite
scores among incoming enrollees, percentage of part-time undergraduates,
student-to-faculty ratios in engineering schools and institution-wide, and
institutional control (public vs. private). Since engineering resources may be
related to institutional size, we also included IPEDS figures for undergraduate, graduate, and total enrollment. Among the possible undergraduate
financial aid measures in IPEDS (e.g., percentage of students receiving aid
from institutions, aid from state-based sources, etc.), we included federal aid
and student loan aid to capture the relationship between students' socioeconomic characteristics and institutional selectivity, revenue, and expenditures
found in the literature (Astin 1993; Astin and Oseguera 2004; Reardon,
Baker, and Klasik 2012; Titus 2006).

Two additional sets of variables measure the proportional size of undergraduate and graduate engineering programs of the APPLES institutions
(tested as exploratory measures of resources), and the process by which
undergraduates declare an engineering major at each school. The "major
declaration process" has been noted as a possible influence on students'
perceptions of and movement into/out of engineering (Garrison et al. 2007;
Lichtenstein et al. 2009).

We obtained the institutional characteristics in the Level 2 model sequence
as aggregates from student reports to the level of the institution and/or
as institution averages as follows (with the source of each measure in parentheses):

SET 1: Academic selectivity (IPEDS). SAT composite, calculated as the sum of SAT-Math 75th percentile score and SAT-Critical Reading 75th percentile score; SAT ratio, calculated as a ratio of SAT-Math 75th percentile score to SAT-Critical Reading 75th percentile score.

SET 2: Percentage of undergraduates receiving financial aid (IPEDS). Percentage of undergraduates receiving federal grant aid; percentage of undergraduates receiving student loan aid.

SET 3: Percentage of part-time undergraduates (IPEDS).

SET 4: Institutional type: Undergraduate/graduate (IPEDS). Percentage of total students who are graduate students; nearly exclusive undergraduate institution versus institution with graduate students, based on percentage of total students who are graduate students and triangulated with Carnegie Classification: Enrollment Profile and Size and Setting (we classified institutions with less than 10 percent of graduate students among total enrollment as a "nearly exclusive undergraduate institution").

*SET 5: Size of institution and engineering programs: **Total** enrollments (IPEDS)*. Estimated enrollment total; percentage of total students enrolled who are in engineering.

*SET 6: Size of institution and engineering programs: **Undergraduate** enrollments (IPEDS)*. Estimated undergraduate enrollment total; percentage of undergraduate students enrolled who are in engineering.

*SET 7: Size of institution and engineering programs: **Graduate** enrollments (IPEDS)*. Estimated graduate enrollment total; graduate engineering students as a proportion of all engineering students.[11]

SET 8: Student-to-faculty ratio. Student-to-faculty ratio (IPEDS); engineering student-to-faculty ratio, calculated as a ratio of total engineering undergraduates (full time plus part time) to engineering teaching faculty (tenure/tenure track plus nontenure track) (ASEE).

SET 9: Major declaration process (APPLES institutional profiles). Student enters institution in engineering program versus student declares engineering major later, after matriculation.

SET 10: Institutional control (APPLES institutional profiles). Public institution versus private institution.

2.2   Findings on Postgraduation Plans of Engineering Students

The APPLES instrument allowed us to examine postgraduation plans relative to pursuing a job in the labor market and/or attending graduate school in engineering or outside of engineering. Survey respondents indicated their postgraduation plans for work or graduate school in four questions, as illustrated in the top panel of table 2.1. Plans for working as a

---

11. "Percentage of graduate students enrolled who are in engineering" cannot be calculated because two institutions in the sample do not have graduate student enrollments.

practicing engineer are most common (~81 percent) as opposed to plans for obtaining a nonengineering job. Graduate school plans, however, are mixed. Forty-three percent of students indicated plans for attending a graduate program in engineering, and nearly one-third indicated graduate plans in other fields. Clearly, many students indicated some degree of interest in both engineering and nonengineering options.

As described earlier, we created two sets of mutually exclusive categories of students from those four separate questions in order to understand student retention in engineering careers more fully. Examining these exclusive categories in the bottom panel of the table, we see that most junior and senior engineering students have not ruled out future employment or education in both engineering and nonengineering fields. In fact, *less than 30 percent of students are strictly engineering focused in their plans*. Among respondents in the narrower sample, the proportions of students who are engineering focused and have cross-field plans are nearly equal.

Table 2.2 shows that the distributions of men and women across

Table 2.1   **Postgraduation plans among junior and senior engineering majors**
($N = 2,143$)

	Percent marking:				
	Definitely not	Probably not	Maybe	Probably yes	Definitely yes
Engineering job	3.2	6.3	9.5	32.9	48.2
Nonengineering job	11.4	33.1	30.0	19.3	6.2
Engineering graduate school	9.9	18.3	28.9	27.1	15.7
Nonengineering graduate school	18.5	27.2	26.0	19.9	8.4

Combinations of postgraduation plans: Percent who are classified as:
Engineering focused	28.1
Nonengineering focused	6.5
Having "all other plans"	65.4

Combinations of postgraduation plans, refined sample analyses ($n = 1,318$): Percent who are classified as:
Engineering focused	45.7
Nonengineering focused	10.5
Having cross-field plans	43.8

Notes: In "combinations of plans," "engineering-focused students" are those who marked probably/definitely yes to one or both engineering options (job, graduate school), and probably/definitely no to both nonengineering options (job, graduate school). "Nonengineering-focused students" are those who marked probably/definitely yes to one or both nonengineering options, and probably/definitely no to both engineering options. "All other plans" include all other possible response combinations to the four postgraduation survey items. "Cross-field plans" include only those response combinations where students marked probably/definitely yes to at least one engineering option and one nonengineering option.

Table 2.2 Postgraduation plans among junior and senior engineering majors by gender, SES, and major[a]

	All students	Gender		Perceived family income[b]			Major								
		Women	Men	Low income	Middle income	High income	Aero	Bio-x	Chem.	Civil/ Env.	Comp. sci./E.	Elec.	Indus.	Mech.	Other
							Full sample								
N	2,143	658	1,465	516	830	756	98	119	125	302	229	295	167	584	224
Engineering focused	28.1	24.6	29.9[c]	27.7	32.0	24.7[d]	26.5	21.0	20.0	41.7	30.1	31.2	9.0	27.1	29.5[e]
Nonengineering focused	6.5	7.9	5.9	3.9	5.2	9.5	3.1	24.4	12.0	2.6	6.1	3.1	11.4	4.5	7.1
All other plans	65.4	67.5	64.2	68.4	62.8	65.7	70.4	54.6	68.0	55.6	63.8	65.8	79.6	68.5	63.4
							Refined sample								
N	1,318	406	903	306	522	472	56	82	69	194	144	193	86	354	140
Engineering focused	45.7	39.9	48.5[c]	46.7	51.0	39.6[d]	46.4	30.5	36.2	64.9	47.9	47.7	17.4	44.6	47.1[e]
Nonengineering focused	10.5	12.8	9.5	6.5	8.2	15.3	5.4	35.4	21.7	4.1	9.7	4.7	22.1	7.3	11.4
Cross-field plans	43.8	47.3	42.0	46.7	40.8	45.1	48.2	34.1	42.0	30.9	42.4	47.7	60.5	48.0	41.4

Note: Aero = aerospace engineering, Bio-x = bio-x engineering, Chem. = chemical engineering, Civil/Env. = civil/environmental engineering, Comp. sci./E. = computer science/engineering, Elec. = electrical engineering, Indus. = industrial engineering, Mech. = mechanical engineering, Other = other engineering.

[a] In "postgraduation plans," "engineering-focused students" are those who marked probably/definitely yes to one or both engineering options (job, graduate school), and probably/definitely no to both nonengineering options (job, graduate school). "Nonengineering-focused students" are those who marked probably/definitely yes to one or both nonengineering options, and probably/definitely no to both engineering options. "All other plans" include all other possible response combinations to the four postgraduation survey items. "Cross-field plans" include only those response combinations where students marked probably/definitely yes to at least one engineering option and one nonengineering option.

[b] The five categories in this response scale are collapsed to three for these analyses: Low income = "low income" and "lower-middle income"; middle income = "middle income"; high income = "upper-middle income" and "high income."

[c] Women mean engineering focused < men mean engineering focused in full sample at *p* < .05, in refined sample at *p* < .01 (independent sample *t*-test).

[d] Middle income mean engineering focused > high income mean engineering focused in full sample at *p* < .01, in refined sample at *p* < .01 (one-way ANOVA with post hoc pairwise comparisons).

[e] Mean engineering focused varies significantly between majors (overall ANOVA for full sample, *p* < .001; overall ANOVA for refined sample, *p* < .001). See text for additional details.

postgraduation plan categories differ from the overall sample. Women are about 34 percent more likely than men to indicate nonengineering-focused postgraduation plans and 13 percent more likely to indicate cross-field plans.

Table 2.2 also reports the distribution of postgraduation plans by students' perceived family income and undergraduate major. Although the trends are not linear, high-income students tend to have the lowest rates of engineering-focused plans and the highest rates of nonengineering-focused plans, with more than double the rate of nonengineering-focused plans than their low-income peers. With regard to major, civil and environmental engineers are the most focused on engineering jobs and/or graduate study, and the least focused on nonengineering jobs and/or graduate study. At over one-third in the refined sample, bio-x majors are the most focused on nonengineering pathways. Industrial engineers have the largest proportion of students interested in both engineering and nonengineering options and the lowest rates of engineering-focused plans across all major groups.

2.2.1   Level 1 MLR Results—Student Factors

Multinomial logistic regression allows us to examine the characteristics in tables 2.1 and 2.2 and additional factors simultaneously in distinguishing students among categories of postgraduation plans. Table 2.3 shows the results of Level 1 MLR models for the full sample and the narrower sample, with engineering-focused students as the reference group.

The patterns for control variables in both full and refined sample multinomial models are consistent with those in earlier work: intrinsic psychological motivation to study engineering, financial motivation to study engineering, exposure to the engineering profession, and academic involvement in engineering are associated with plans oriented exclusively to engineering. For instance, students scoring one standard deviation higher on intrinsic psychological motivation toward engineering have just .38 times (exp(−.98)) the odds of being nonengineering focused relative to their lower-scoring peers (panel A). In contrast, students scoring one standard deviation higher on participation in nonengineering extracurricular activities and confidence in professional/interpersonal skills have 1.6 and 2.0 times the odds, respectively, of being nonengineering focused.

Note also that for most of the variables in the models, statistically significant coefficients for the full and refined samples are similar in magnitude, as are the standard errors. In a few cases, coefficients for the same variable differ considerably across the two models, which may indicate sensitivity to either sample size or the different compositions of the "all other plans" and "cross-field plans" categories; those results should be interpreted with caution. For example, grade point average (GPA) self-reported by students on the APPLES instrument differentiates students with engineering-focused plans from those with "all other plans," but not from those with specific cross-field plans nor students with nonengineering-focused plans. Supplementary HLM models of each constituent pathway find that GPA

Table 2.3   Results of multinomial logit models of students' postgraduation plans

A. Postgraduation plans: Full sample (N = 2,143)

B. Postgraduation plans: Refined sample (N = 1,318)

	Level-1 conditional model (Full sample)			Level-1 conditional model (Refined sample)		
	Coefficient	SE	P-value	Coefficient	SE	P-value
For category 0: Nonengineering focused (reference group: engineering focused)						
Intercept	-2.50	0.42	0.000	-2.58	0.46	0.000
URM women[a]	-0.95	0.65	0.143	-0.97	0.71	0.173
Perceived family income (standardized)	0.07	0.12	0.561	0.03	0.13	0.826
Financial motivation (standardized)	-0.49	0.11	0.000	-0.57	0.12	0.000
Intrinsic psychological motivation (standardized)	-0.98	0.10	0.000	-1.11	0.12	0.000
Exposure to engineering work (standardized)	-0.38	0.12	0.002	-0.49	0.13	0.000
Involvement in engineering classes (standardized)	-0.39	0.11	0.001	-0.40	0.13	0.002
Participation in nonengineering activities (standardized)	0.46	0.14	0.002	0.52	0.15	0.001
GPA index (standardized)	0.08	0.12	0.512	0.10	0.14	0.485
Professional/interpersonal confidence (standardized)	0.68	0.12	0.000	0.73	0.13	0.000
Aerospace engineering[b]	-0.43	0.77	0.577	-0.47	0.83	0.573
Bio-x engineering[b]	1.44	0.42	0.001	1.62	0.46	0.001
Chemical engineering[b]	1.19	0.47	0.012	1.58	0.52	0.003
Civil/environmental engineering[b]	-1.31	0.48	0.006	-1.15	0.52	0.026
Computer science/engineering[b]	0.67	0.47	0.154	0.92	0.50	0.067
Electrical engineering[b]	0.17	0.45	0.701	0.16	0.47	0.734
Industrial engineering[b]	0.78	0.49	0.108	0.53	0.53	0.323
Other engineering[b]	0.32	0.48	0.503	0.34	0.53	0.514
Expected salary given major field and state (using median salaries from BLS occupation data, 2007) (standardized)	-0.34	0.17	0.044	-0.35	0.18	0.057

(continued)

Table 2.3   (continued)

	A. Postgraduation plans: Full sample (N = 2,143)			B. Postgraduation plans: Refined sample (N = 1,318)		
	Level-1 conditional model			Level-1 conditional model		
	Coefficient	SE	P-value	Coefficient	SE	P-value
For category 1: All other plans (reference group: engineering focused) / Cross-field plans (reference group: engineering focused)						
Intercept	0.97	0.15	0.000	0.03	0.18	0.870
URM women[a]	0.28	0.22	0.205	0.66	0.26	0.011
Perceived family income (standardized)	−0.02	0.06	0.692	0.00	0.07	0.957
Financial motivation (standardized)	−0.03	0.05	0.596	0.02	0.07	0.762
Intrinsic psychological motivation (standardized)	−0.30	0.06	0.000	−0.29	0.08	0.000
Exposure to engineering work (standardized)	−0.06	0.06	0.302	−0.07	0.07	0.337
Involvement in engineering classes (standardized)	−0.27	0.06	0.000	−0.35	0.07	0.000
Participation in nonengineering activities (standardized)	0.05	0.06	0.369	0.08	0.07	0.252
GPA index (standardized)	−0.18	0.06	0.002	−0.11	0.07	0.124
Professional/interpersonal confidence (standardized)	0.26	0.06	0.000	0.46	0.07	0.000
Aerospace engineering[b]	0.05	0.29	0.859	−0.01	0.36	0.982
Bio-x engineering[b]	−0.05	0.28	0.864	0.16	0.33	0.636
Chemical engineering[b]	0.27	0.27	0.327	0.31	0.34	0.365
Civil/environmental engineering[b]	−0.73	0.18	0.000	−0.71	0.23	0.002
Computer science/engineering[b]	−0.06	0.23	0.801	0.01	0.28	0.968
Electrical engineering[b]	−0.07	0.19	0.701	−0.01	0.23	0.969
Industrial engineering[b]	0.67	0.30	0.027	0.54	0.34	0.119
Other engineering[b]	−0.20	0.24	0.403	−0.11	0.30	0.720
Expected salary given major field and state (using median salaries from BLS occupation data, 2007) (standardized)	−0.09	0.10	0.374	−0.07	0.12	0.534
Variance components						
Tau category 0	2.84			3.40		
Tau category 1	0.40			0.52		
Sigma squared[c]	3.29			3.29		

Note: Robust standard errors could not be computed for the level-1 conditional models. Preliminary analyses were conducted testing URM status and gender as separate measures. Coefficients for both were nonsignificant in the full sample model. In the refined sample model, URM status was a positive predictor of category 1 ("cross-field plans") ($b = .42$, $se = .19$, $p = .028$), and gender ($0 =$ female, $1 =$ male) was borderline significant and negative ($b = −.30$, $se = .15$, $p = .051$).

[a]Reference group: All other students.

[b]Reference group: Mechanical engineering.

[c]Approximation based on the standard logistic distribution where variance equals pi squared/3.

is a negative predictor of plans to pursue an engineering job ($b = -.59$, $p < .001$) and a positive predictor of plans to pursue engineering graduate school ($b = .20, p < .001$) with no significant relationship between GPA and plans for nonengineering jobs or graduate study. At a minimum, these results suggest that students with higher grades have a comparable likelihood of being nonengineering and engineering focused at this stage, although engineering focus may entail an engineering graduate degree rather than engineering employment.

With these factors statistically controlled, we focus on relationships between postgraduation plans and SES, major, and salary by field and state where the student was educated.[12] We give particular attention to URM women in engineering on the basis of earlier analyses. The second panel of table 2.3 shows that being a URM woman increases the odds of having cross-field plans versus engineering-focused plans in the refined sample. Supplementary HLM analyses indicate that URM women are more likely than their peers to be considering nonengineering jobs ($b = .22, p < .05$) and engineering graduate school ($b = .22, p < .05$). Although our perceived family income SES measure is negatively correlated with plans for pursuing an engineering job and an engineering graduate degree ($r = -.13$ and $r = -.14$, respectively, $p < .001$), a finding that holds in the supplementary models ($b = -.56$ and $b = -.10$, respectively, $p < .001$), it does not distinguish engineering-focused students from others in the MLR analysis, indicating that it works through other variables in the MLR model.

Relative to students in mechanical engineering, civil/environmental engineering majors are more likely to report strictly engineering-focused career plans. Students who are bio-x and chemical engineering majors, on the other hand, have three to five times the odds of having nonengineering plans as do mechanical engineering majors. Industrial engineers (full sample only) tend to report mixed plans. Students' plans in other majors are statistically similar to those among mechanical engineering majors.

Labor market influences were measured by field-specific median earnings in the state where a respondent's university is located. Table 2.4 shows that across all states in our sample, earnings vary considerably with major, being highest for aerospace and computer engineers and lowest for biomedical and civil engineers.[13] In our full-sample MLR model, higher salaries differentiated engineering-focused students from nonengineering-focused ones.

2.2.2   Level 2 MLR Results—Institutional Factors

At Level 2, we tested for institutional factors related to students' pathways in or outside of engineering. Because of the limited number of institutions in our sample, we examined seventeen institutional characteristics in ten

12. Consideration of labor market earnings and majors distinguishes our Level 1 analysis from the earlier APPLES project (Sheppard et al. 2010).
13. Table 2.4 and national estimates differ due to state-by-state variation in field specific labor markets (see BLS 2011).

Table 2.4   May 2007 average median salary among states in the APPLES
institutional sample by occupational category

Occupational category	Average median salary
Aerospace engineers	82,847
Computer hardware/software engineers[a]	82,316
Chemical engineers	77,842
Electrical engineers	77,455
Industrial engineers	70,630
Mechanical engineers	70,219
Environmental engineers	69,637
Civil engineers	68,598
Biomedical engineers	67,299

Source: Bureau of Labor Statistics (2011).

Notes: These occupational categories map to the "core" APPLES majors in the present study. Average median salaries for other occupational categories that map to smaller APPLES majors (e.g., "marine engineers") are available on request.

[a]This category represents the aggregate of three BLS categories: computer hardware engineers; computer software engineers, applications; and computer software engineers, systems software.

separate models. Figure 2.1 captures statistically significant ($p < .05$) results and includes estimated coefficients for the full sample only. Because of the small number of institutions in the APPLES sample, results should be interpreted with caution.

Relative to engineering-focused students, nonengineering-focused students are more likely to attend institutions that enroll undergraduate populations with higher SAT scores, lower rates of federal financial aid receipt, and lower rates of part-time attendance. Students attending private institutions also are more likely to be nonengineering focused (> ten times the odds) than their counterparts at public institutions. The magnitudes of the coefficients for the standardized variables suggest that institution-level SES (as reflected by the receipt of aid) is the strongest differentiator of postgraduation plans. We note that several of these institutional characteristics are intercorrelated, with the strongest correlation observed between SAT composite score and private institutions ($r = .81, p < .001$) and SAT composite score and percentage of students receiving federal aid ($r = -.78, p < .001$).

The size and type of programs did not differentiate postgraduation plans, with sole exception of graduate student enrollment: undergraduate engineering majors attending institutions with higher proportions of graduate students tend to have higher rates of nonengineering postgraduation plans (in the APPLES sample of schools, the proportion of graduate students is not correlated with other significant institution-level predictors). Institutional characteristics did not differentiate students with engineering-focused plans from those with "all other plans" or more specific "cross-field plans"; and whether an engineering major is declared upon admission or

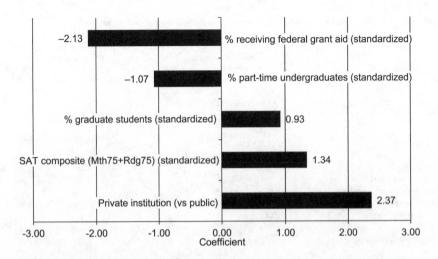

Fig. 2.1   **Level 2 predictors of nonengineering-focused plans (vs. engineering-focused plans), full sample analysis (*N* = 2,143)**

Note: Predictors are significant at *p* < .05 holding all Level 1 variables constant. Due to limitations of institutional sample size (*N* = 21), each variable was tested in a separate model. Standard errors are available on request. Methodology reports additional details.

during matriculation is not related to plans to persist in engineering post-baccalaureate.

2.2.3   Exploring the Role of SES

Supplementary HLM analyses of postgraduation plans indicate a negative relationship between perceived family income and plans to work or to study in an engineering field, holding all other variables in the model constant. Since SES does not emerge as significant in the MLR analyses (table 2.3), we conducted additional analyses to better understand its role in the postgraduation plans of engineering students. Table 2.5 shows how low-, middle-, and high-income students differ by major and by institutional characteristics related to plans in the MLR analyses. Patterns by major fit with the results in tables 2.2 and 2.3. Modestly greater proportions of high-income students than low-income students major in fields associated with nonengineering postgraduation plans (bio-x, chemical, and industrial) and a somewhat smaller proportion major in civil/environmental engineering, which is associated with engineering-focused plans. These differences are not significantly different from zero, however.

With respect to institutional factors that might be related to SES, we earlier found that more selective and private institutions were associated with nonengineering-focused plans. Table 2.5 shows the expected trend between SES and institutional selectivity and SES and institutional control, with low-income students 56 percent less likely than high-income students to

Table 2.5   **Student SES by major and by select institutional characteristics[a]**

	Percentage among			Significance tests (one-way ANOVA post hoc pairwise comparisons)[b]		
	Low income	Middle income	High income	Mean low inc. vs. mean mid. inc.	Mean mid. inc. vs. mean high inc.	Mean low inc. vs. mean high inc.
N	*516*	*830*	*756*			
		Major				
Aerospace engineering	3.9	5.1	4.4	n/s	n/s	n/s
Bio-x engineering	4.1	4.9	6.9	n/s	n/s	n/s
Chemical engineering	5.6	5.4	6.3	n/s	n/s	n/s
Civil/environmental engineering	15.3	15.3	12.0	n/s	n/s	n/s
Computer science/engineering	13.4	10.4	9.7	n/s	n/s	n/s
Electrical engineering	14.9	13.1	13.5	n/s	n/s	n/s
Industrial engineering	6.4	6.7	9.7	n/s	n/s	n/s
Mechanical engineering	27.3	29.2	25.5	n/s	n/s	n/s
Other engineering	9.1	9.9	12.0	n/s	n/s	n/s
		Institutional characteristics				
SAT composite (math 75th + reading 75th): Quartiles						
Highest quartile (1,500–1,580)	12.4	13.9	28.0	n/s	***	***
Second-highest quartile (1,290–1,460)	36.6	39.3	40.5	—	—	—
Third-highest quartile (1,130–1,250)	32.8	32.2	24.5	—	—	—
Lowest quartile (960–1,100)	18.2	14.7	7.0	n/s	***	***
Institutional control[c]						
Public institution	70.7	71.1	58.3	—	—	—
Private institution	29.3	28.9	41.7	n/s	***	***
Percent undergraduates receiving federal financial aid: Quartiles						
Highest quartile (37–69%)	21.7	12.2	6.2	***	***	***
Second-highest quartile (22–28%)	20.2	20.0	14.6	—	—	—
Third-highest quartile (17–21%)	47.7	54.7	54.8	—	—	—
Lowest quartile (6–13%)	10.5	13.1	24.5	n/s	***	***
Percent part-time undergraduates: Quartiles						
Highest quartile (28–44%)	26.7	18.8	8.5	**	***	***
Second-highest quartile (16–26%)	32.6	35.9	31.5	—	—	—
Third-highest quartile (7–12%)	17.4	23.4	30.8	—	—	—
Lowest quartile (0–3%)	23.3	21.9	29.2	n/s	**	*
Percent graduate students: Quartiles						
Highest quartile (29–60%)	33.1	39.8	48.3	*	**	***
Second-highest quartile (18–26%)	41.9	39.3	31.9	—	—	—
Third-highest quartile (15–17%)	16.7	13.7	8.2	—	—	—
Lowest quartile (0–12%)	8.3	7.2	11.6	n/s	**	n/s

[a]The five categories in the SES response scale are collapsed to three for these analyses: Low income = "low income" and "lower-middle income"; middle income = "middle income"; high income = "upper-middle income" and "high income."

[b]For institutional characteristics cut into quartiles, ANOVAs were conducted among the highest quartile and lowest quartile only.

[c]Of the twenty-one institutions in the APPLES institutional sample, thirteen are public and eight are private.

***$p < .001$

**$p < .01$

*$p < .05$

n/s = not significant

attend the most selective schools, and over two times more likely to attend the least selective institutions. High-income students are 42 percent more likely to attend a private institution. Qualities of the student body that predict engineering-focused plans are also associated with SES. Low-income students are more likely than high-income students to attend high financial aid institutions and schools with greater part-time attendance and less likely to attend schools with high percentages of graduate student enrollments. These patterns suggest that institutions play a role in the differing postgraduation destinations of low- and high-SES engineers.

2.2.4   Exploring the Role of the Labor Market

Table 2.3 showed that the median salary of professionals in the same field and in the same state in which students' college or university is located modestly differentiates students with engineering-focused postgraduation plans from students with nonengineering-focused plans. In supplementary HLM analyses, the earnings variable does not emerge as a significant predictor of constituent pathways, although it is weakly correlated with plans to pursue engineering graduate study ($r = .13$, $p < .001$) and to obtain an engineering job ($r = -.05$, $p < .05$). A closer look at findings by major sheds light on this pattern. Relative to other majors, computer science/engineering majors have the highest rate of intention to pursue engineering graduate school (51 percent reporting "probably" or "definitely" yes; data available by request), and computer hardware/software engineers are among those with the highest earnings (table 2.4). This helps explain the disappearance of the positive simple correlation between earnings and engineering graduate school when major is controlled in the supplementary model. Moreover, in the companion model of plans to pursue an engineering job, computer science/engineering majors are significantly less likely to consider engineering jobs than mechanical engineering majors (the reference group). The negative simple correlation between salary and engineering job plans could thus be an artifact of this computer science/engineering trend.

The modest role of the salary measure in these models does not mean that financial motivation is irrelevant to students. Financial motivation to pursue engineering studies differentiates students with engineering-focused plans from those focused on nonengineering options (table 2.3); in the HLM models of constituent pathways, it is a positive predictor of plans to work in an engineering job after graduation ($b = .93$, $p < .001$). Interestingly, there is no difference in mean financial motivation by students' perceived family income.

2.2.5   Exploring the Role of Major Field of Study

Delving deeper into the "cross-field plans" category in table 2.2 allows us to analyze further the career plans among students who are neither engineering nor nonengineering focused—students who indicate strong interest in pursuing at least one of two engineering pathways (job or graduate school) and at least one of two nonengineering pathways (job or graduate school). The proportion of students with an engineering focus varies from 17 percent (industrial engineering majors) to 65 percent (civil and environmental majors), whereas students with a nonengineering focus varies from 4 percent (civil and environmental majors) to 35 percent (bio-x majors). Civil and environmental engineering majors are least likely to have cross-field plans (30.9 percent), and industrial engineering majors are most likely (60.5 percent).

The analysis in table 2.6 is framed around two clusters of majors: those with 40 percent or more who are engineering focused (aerospace, civil/environmental, computer science/engineering, mechanical, electrical, and other engineering) and those with less than 40 percent who are engineering focused (bio-x, chemical, industrial). Note that the motivation and confidence means are calculated among the full sample of respondents even as the clusters are demarcated by refined sample proportions. We highlight patterns in civil/environmental, bio-x, and industrial engineering to exemplify interrelationships.

The cluster of majors with 40 percent or more of engineering-focused students includes some of the oldest fields in engineering, as indicated by the year the fields' professional societies were established: the American Society of Civil Engineering (1852), the American Society of Mechanical Engineers (1880), and the Institute of Electrical and Electronic Engineers (1884). Civil engineering majors are the most focused in their engineering plans. They also have higher mean scores on intrinsic psychological motivation and lower mean scores on financial motivation (which fits with their relatively lower earnings in table 2.4), and among the lowest mean scores for professional/interpersonal confidence "as compared with [their] classmates." This may reflect selection effects. The MLR model coefficients also indicate that even with these motivation and confidence measures controlled, civil/environmental majors have significantly higher odds of engineering-focused plans than students in the reference group, mechanical engineering (table 2.3).

In the second cluster, less than 40 percent of students have engineering-focused plans. This cluster contains two of the more recently created fields, industrial engineering and bioengineering (the Institute of Industrial Engineers was established in 1948, and the American Institute for Medical and Biological Engineering was established in 1991). Bio-x majors, in fact, have the highest proportion of students focused on nonengineering graduate school and/or job options after college. Simultaneously, they report significantly lower financial motivation to study engineering than students in any other major group except for aerospace, civil/environmental, and other engineering, lower intrinsic psychological motivation, and average levels of professional/interpersonal confidence. Perhaps many bio-x students select

Table 2.6   Mean construct scores by major and gender: Intrinsic psychological motivation, financial motivation, and professional/interpersonal confidence

	Intrinsic psychological motivation			Significance test for gender difference (t-test)	Financial motivation			Significance test for gender difference (t-test)	Professional/interpersonal confidence			Significance test for gender difference (t-test)
	All students	Women	Men		All students	Women	Men		All students	Women	Men	
Higher rates of engineering-focused plans[a]												
Aerospace engineering	82.5	87.2	81.8	n/s	58.9	45.3	61.0	n/s	68.5	64.1	69.1	n/s
Civil/environmental engineering	80.8	82.9	79.7	n/s	64.4	62.8	65.4	n/s	67.0	67.7	66.7	n/s
Computer science/engineering	80.5	75.0	81.9	n/s	66.0	65.2	65.9	n/s	66.7	66.8	66.6	n/s
Electrical engineering	82.8	87.7	81.5	*	67.5	72.3	66.2	n/s	69.3	72.1	68.7	n/s
Mechanical engineering	81.4	78.9	82.1	n/s	64.6	65.1	64.8	n/s	68.1	68.9	67.9	n/s
Other engineering	81.5	83.4	79.6	n/s	63.8	61.9	65.5	n/s	68.9	66.6	70.9	n/s
Lower rates of engineering-focused plans[a]												
Bio-x engineering	78.2	77.4	78.7	n/s	55.8	52.6	58.8	n/s	69.4	66.4	72.2	*
Chemical engineering	77.3	75.6	79.0	n/s	66.3	68.5	64.5	n/s	68.9	67.5	70.8	n/s
Industrial engineering	69.7	66.1	73.2	n/s	71.5	73.8	68.6	n/s	73.3	70.5	76.1	*

Notes: Among all students, ANOVA for intrinsic psychological motivation is significant at $p < .001$. Looking at pairwise comparisons, all means are significantly different from industrial engineering except for chemical engineering. Among all students, ANOVA for financial motivation is significant at $p < .001$. Looking at pairwise comparisons, all means are significantly different except from bio-x engineering except for aerospace engineering, civil/environmental, and other engineering. In addition, aerospace engineering is significantly different from industrial engineering. Among all students, ANOVA for professional/interpersonal confidence is significant at $p < .01$. Looking at pairwise comparisons, industrial engineering is significantly different from civil/environmental engineering, computer science/engineering, and mechanical engineering.

[a] "Higher rates of engineering-focused plans" include those majors with 40 percent or more of students who report engineering focus, based on the refined sample proportions (see table 2.2). "Lower rates" include those majors with less than 40 percent. Means are reported among students in the full sample.

*** $p < .001$
** $p < .01$
* $p < .05$
n/s = not significant

their majors as a bridge to other industries and occupations, not for the affective and financial returns to engineering studies per se.

Despite the comparable (and recent) histories of the respective fields, bio-x majors contrast with industrial engineering majors, who have lower rates of engineering and nonengineering-focused plans, and higher rates of cross-field plans. Table 2.6 shows that industrial engineering majors: (a) are significantly less psychologically motivated to study engineering than students in every other major group except chemical engineering; (b) have the highest rates of financial motivation to study engineering; and (c) have the highest professional/interpersonal confidence, significantly higher than students in some of the largest (and oldest) degree programs in engineering (mechanical engineering and civil engineering). These data suggest that students who pursue industrial engineering degrees may have a distinctive set of interests and self-concepts, and that industrial engineering academic environments may have unique emphases relative to other engineering fields, which we explore in the next section.

Motivation and confidence does not differ much between women and men within major in the APPLES sample, with some variation on some dimensions (e.g., in industrial engineering men are neither more nor less financially motivated to study engineering than are women, but are more confident in their professional/interpersonal skills). Motivational differences between majors are more pronounced among women than among men, while confidence differences between majors are more pronounced among men than among women.[14] Recent work suggests that women may pursue engineering degrees for different reasons than men (Orr et al. 2009), which might help explain the different proportions of women and men in engineering subfields to begin with (NSF 2011a).

2.2.6   The Dynamic Early Career Path of Engineers

The data on late-stage engineering majors (juniors and seniors) opens a window into how today's engineering students conceive their professional futures, and how educational and employment contexts may affect their plans. *The most striking finding in our analysis in this regard is that a substantial majority of engineering students are not committed to a purely engineering future.* That over two-thirds of students in the APPLES sample have nonengineering, mixed, or uncertain plans raises questions about what drives students to pursue pathways outside of engineering. One conjecture is that the engineering major and field communicate a flexible career pathway to students, particularly in certain subfields. Perhaps the training students receive in engineering programs illuminates options through and outside of

---

14. When data are disaggregated by gender, industrial engineering majors have the highest mean professional/interpersonal confidence among men only, suggesting stronger self-selection based on confidence for men, conditional effects of the industrial engineering academic environment, or both.

traditional engineering. A more negative interpretation is that experience in the major leads many students to reconsider their engineering career plans. While we cannot differentiate precisely between these interpretations, additional findings are suggestive of the former.

First, nonengineering-focused students tend to be lower in intrinsic motivation for engineering work and its financial rewards—an indication that the field itself is not a strong draw for a subset of students who nonetheless persist in the major to their third and fourth year of college. Further, these students have relatively high professional/interpersonal self-confidence and tend to attend private, selective, and well-funded institutions. They thus appear to be among the better-prepared and competitive undergraduate engineers whose college experiences arguably give them a broader view of viable career options for someone with their training and skill set. This interpretation is supported by Lowell et al. (2009), who document a significant decline in the retention of top academic performers in science, technology, engineering, and mathematics (STEM) related jobs after college, and by Herrera and Hurtado (2011), who show that college students at more selective institutions have a lower probability of sustained STEM career interests than students at less selective institutions.

The APPLES data measure the point at which students are starting to think tangibly about their professional futures along engineering career pathways. To add further perspective to the findings above, we turn to data from the 2006 National Survey of Recent College Graduates (NSRCG) (NSF 2011b), as analyzed and reported in Sheppard et al. (2014). These data, collected from graduates of 2003, 2004, and 2005, show that some two years after graduation, 60 percent of engineering graduates were working in an engineering job, 18 percent were working in other science and engineering-related fields,[15] and 14 percent were working in nonengineering-related fields (the balance were in school or not employed). Compared with the approximately one-third of APPLES junior and senior engineering majors reporting engineering-focused postgraduation plans, the NSRCG data raise the possibility that the early career pathway of engineers may still be in a highly dynamic stage. Indeed, the 32 percent of recent graduates working outside of engineering[16] exceeds the 26 percent of students in the APPLES sample who planned to pursue nonengineering jobs after graduation, and is considerably higher than the 7 percent who were exclusively focused on a nonengineering pathway. The primary reasons that engineering graduates gave for pursuing jobs "unrelated to their highest degree" provide insight into why a majority of engineering students are not planning for engineering options only following graduation. These include "pay and

15. Computer and mathematical sciences, life sciences, physical sciences, social sciences, and "other" related fields.
16. Of whom 44 percent (= 14/32) were in fields not related to engineering and science.

promotion opportunities," "working conditions," "job location," "change in professional/career interests," "family," and "job in highest degree field not available" (Sheppard et al. 2014).

If the career pathways of today's young engineers are marked by flexibility and flux, how does this compare with engineering pathways in the past? Earlier NSRCG data indicate that trends are not substantially different. Findings from the 1993 NSRCG, which surveyed graduates of 1991 and 1992, show that 61 percent of engineering graduates were employed in an engineering occupation (NSF 2012). This suggests that engineering graduates are neither more nor less likely to pursue engineering work now than they were in the 1990s. Slightly larger shifts are seen in other occupations. Recent engineering graduates are more likely than 1990s graduates to pursue science occupations and computer- and math-related occupations, and less likely to pursue occupations unrelated to science and engineering. Thus, while engineering graduates have long applied their degrees to nonengineering work, the destination occupations are variable.

To put these student and practitioner findings on engineering pathways in a larger context, consider professional persistence in other fields. Other professions vary in rates of persistence to practice upon degree completion (where "degree" refers to the minimum credential to practice—a bachelor's degree in engineering compared with an MD for medicine and a JD for law). The pattern of persistence into the profession among engineering graduates with a BS is closer to that in law, where about two-thirds of U.S. law graduates from 2011 obtained a job requiring passage of the bar (National Association for Law Placement 2012a, 2012b), than with patterns in medicine, where the vast majority of graduates pursue additional medical-related education after obtaining their MD (National Resident Matching Program and the Association of American Medical Colleges 2011).[17] Although the dynamics of self-selection into or out of a profession may differ depending on the profession and its licensure processes, these statistics are notable in contextualizing engineering's migratory pathways after investing in the "gateway" degree.

17. Among the 41,623 U.S. law graduates from the class of 2011 for whom employment status was known (94 percent of the entire class), 65.4 percent obtained a job requiring passage of the bar. An additional 12.5 percent reported having jobs for which a law degree is beneficial, but passage of the bar is not required. About 7 percent were employed in a nonprofessional or a professional position other than law, and less than 1 percent was employed in an unknown position. The remaining 15 percent were not employed or full-time students (National Association for Law Placement 2012a, 2012b). In contrast, the total number of medical school graduates in 2011 was 17,364 (Association of American Medical Colleges 2011). Fully 16,599 of the active applicants in the 2011 Main Residency Match were seniors from U.S. Allopathic Medical Schools (National Resident Matching Program and the Association of American Medical Colleges 2011). This suggests that the overwhelming majority (over 95 percent) of MDs go on to seek additional training to practice medicine.

2.2.7   The Role of Engineering Subfields

Postgraduation plans, as well as individual characteristics such as motivation for pursuing engineering studies and professional/interpersonal self-confidence, vary by engineering major. Early career occupational outcomes vary similarly. For example, consistent with our results on plans, civil engineering majors in the NSRCG 2006 data set reported the highest rates of working in the same field as their major two years after college (Sheppard et al. 2014). It appears that some majors are more tightly coupled with engineering career pathways than are others. Individual characteristics of students in the major may explain at least part of this coupling.

Individual characteristics may have both a direct relationship with postgraduation plans and an indirect relationship, operating through self-selection into certain fields. For instance, students choosing civil engineering may perceive technical career pathways that appeal to their enjoyment of engineering work. Bio-x students, with lower levels of financial and psychological motivation to study engineering, may select their programs as a route to medical school pathways. Students with high levels of financial motivation may be drawn to industrial engineering programs because they perceive hybrid (engineering and management) pathways through them that yield higher financial returns. This interpretation assumes that the effects of major are incidental to the institution in that students take cues from the field and translate them into decisions of major choice. However, Brawner et al. (2009, 2012) found that the mode by which students enter an engineering major may have some association with the major they choose. For example, students declaring industrial engineering were more likely to have come to this major from an "undecided" status within schools of engineering as compared with "direct admits" and students enrolled in schools with mandatory first-year engineering programs. Moreover, industrial engineering was among the few engineering programs that gained student majors after the third semester of college and through graduation. Recall that the major declaration process did not show a statistically significant relationship with students' postgraduation plans in the present study; Brawner et al.'s findings hint that the effect of matriculation channels may act indirectly through major choice.

A different interpretation assumes that student experiences in the major influence postgraduation choices. Cultural distinctions between engineering fields can be present among various departments of engineering in a college, thereby exposing students to distinctive values, norms, and expectations pertaining to their major. In her study of five engineering departments, Petrides (1996) found departmental cultural characteristics (e.g., prestige of the field, what motivated people to be in the field, percentage of women in the field) to be related to graduate students' postgraduation plans. This suggests that different branches of engineering attract and socialize individuals in different ways and provide distinct and varied models of the school-to-career connection. We might further expect that specific engineering departments vary in important ways in their pedagogy, curriculum, and other educational experiences that give rise to variation in postgraduation plans and expectations across the subfields. Given how career plans vary with the age of subfields, broad cultural differences in the expectations for career outcomes would not be surprising.

Other drivers include field- and region-specific labor markets—the data in this study indicate some sensitivity to the salaries in each engineering-occupational category. However, it may be that the stronger driver is the alternative salaries available in nonengineering positions and industries connected to students' majors (transmitted through socialization, job fairs, internship opportunities, alumni contacts, and so on). These alternative salary data for students in each major are not collected in a systematic, cross-institutional way, although records at individual institutions' career development offices might offer preliminary insight.

2.2.8   Additional Student-Level Factors in Engineering and Nonengineering Pathways

Findings from the multinomial models show that URM women are more likely to have cross-field postgraduation plans than their peers, while supplementary models find that URM women are more likely to be looking toward engineering graduate school and nonengineering jobs.[18] Sheppard et al.'s (2010) analyses of senior respondents in the APPLES data indicated that URM and non-URM women and men have comparable rates of exposure to the engineering profession, interaction with instructors, and involvement in engineering courses. At the same time, URM women and men ascribed more importance to professional/interpersonal skills in engineering practice, and were more psychologically motivated to study engineering than their non-URM peers. Senior URM women also reported significantly lower grade averages than did senior non-URM women (the same was not true among senior men and among first-year URM and non-URM women in the APPLES sample). Thus, different groups of students may have similar rates of participation in various aspects of their engineering programs, but different kinds of motivation for pursuing engineering work. Other researchers have noted that academic and background characteristics operate differently for URM and non-URM students in terms of career pathways. For example, in longitudinal models of sustained STEM-career interests, Herrera and Hurtado (2011) found that positive predictors for URM students included

18. URM women are 6.8 percent of the sample. Nationally, they represent only 3.7 percent of all enrolled engineering students, 3.1 percent of all engineering bachelor's degree earners, and 1.8 percent of all employed engineers (NSF 2011a). Their experiences in engineering education merit particular attention in terms of the supply and diversity of engineering professionals in the coming decades.

high school GPA and working with a faculty member on research, while positive predictors for non-URM students included SAT score (and negative predictors included SES). Of course, which specific experiences encourage URM women to consider cross-field career pathways more so than other students are not identified in our study.

Exposure to Professional Engineering Work

Building on work that identifies career-planning correlates of internships (Margolis and Kotys-Schwartz 2009), the findings from these models suggest that exposure to a professional engineering environment through internships, visits, and employment increases the odds of having engineering-focused plans versus nonengineering-focused plans. The causal order of this relationship is unclear, but the link is important to conversations about school-to-work opportunities in the undergraduate engineering curriculum. In a study of 484 alumni from four U.S. engineering schools, Brunhaver et al. (2012) found that alumni employed in engineering fields four years after graduating were more likely to have participated in an internship or co-op as an undergraduate than were alumni who were employed in nonengineering fields. They also were more likely to report that they had been hired to an employed position through an undergraduate internship or co-op. This suggests that undergraduate internship and co-op opportunities may help to build a bridge for college students to engineering career pathways, although it is not evident which parts of these experiences reinforce retention in the field besides the possibility of a firm job offer.

Professional/Interpersonal Confidence

The multivariate models show that engineering students with higher professional/interpersonal confidence are less focused on engineering pathways, and more focused on nonengineering or cross-field pathways, after graduation. This finding relates to Salzman and Lynn's (2010) research on industry perceptions of new engineering hires that suggests that companies see new hires as not lacking for technical skill, but as having lower than desired levels of communication and business skills. Managers stressed that engineers must have not only technical competence, but the ability to articulate ideas and collaborate across business functions in order to be successful. Our models hint that engineers with these abilities may be eyeing nonengineering opportunities from the get-go; it is possible that students' participation in professional engineering environments as undergraduates fails to expose them to the full range of skills and talents needed among engineers, and/or to show them how these skills are applied in these settings. It also is possible that students with nonengineering plans and higher levels of professional/interpersonal confidence opt out of those undergraduate experiences altogether, thus limiting their view on the relevance of diverse skills in engineering practice.

However, Brunhaver et al. (2013) found that engineering majors employed in engineering four years after graduating have levels of professional/interpersonal self-efficacy comparable to engineering majors in nonengineering employment. This suggests that time in postgraduation employment may compensate for initial differences in confidence. It should be noted that APPLES confidence measures asked students to rate themselves relative to classmates, while the professional self-efficacy measures referred to task-specific confidence in, for example, "communicating my ideas effectively to people in different positions or fields."

Other data highlight the role of factors outside the college environment that influence professional/interpersonal confidence. Sheppard et al. (2010) observed that both first-year and senior engineering students who reported higher family income levels had higher professional/interpersonal confidence than peers from lower family income backgrounds. Additionally, Colbeck, Campbell, and Bjorklund's (2000) qualitative evaluation of group work among undergraduate engineers reported that while students gained appreciation for communication and conflict resolution in project design teams—skills they perceived as important to professional practice—they did not see faculty providing much guidance in how to work effectively in these teams. Students drew from other experiences (inclusive of internships) to develop understanding of what group work meant and looked like. This raises questions of which curricular components link professional/interpersonal skills to engineering work in a systematic, developmental, and structured way.

Socioeconomic Background

We also note that socioeconomic status appears to be playing a role in the pathways to engineering-focused careers. Undoubtedly, one of the pull factors into the engineering profession is the promise of one of the better immediate returns on education in the form of salaried compensation (Carnevale, Rose, and Cheah 2011; Carnevale, Strohl, and Melton 2011). It is not clear if lower SES students see a career in engineering as a means to realize greater returns on their education more so than do higher SES students.[19] We do find that lower SES students tend to matriculate at less selective, public institutions and institutions that enroll greater proportions of part-time students and students receiving financial aid (a finding supported by Astin [1993], Astin and Oseguera [2004], Reardon, Baker, and Klasik [2012], and Titus [2006]). The results linking these institutional characteristics to engineering-focused students imply a peer effect at such institutions that favors engineering-focused pathways. While public and less selective insti-

---

19. Previous research suggests that lower SES students have similar perceptions of college returns as do their higher SES peers (Rouse 2004) and have similar returns to a bachelor's degree versus a high school diploma (Perna 2003), but this research has not been conducted specifically on engineering students.

tutions may not be intentionally creating norms around engineering career pathways, such norms may be developed by the students (who are more often lower income) they tend to attract and enroll. More focused study of the normative expectations for postgraduation pathways at more/less selective colleges, public/private institutions, and more generally those that enroll primarily lower/higher SES students is needed to better understand these relationships.

2.3 Implications

The findings that most engineering majors see engineering jobs in their futures, but not to the exclusion of nonengineering plans, and that students with engineering-focused plans have distinctive profiles connected with individual factors and their major and institutional characteristics have implications for educational practice.

First, it suggests that engineering education needs to build more awareness of the dynamic pathways of current and future students through informal and formal programs that allow students to see different applications of engineering work.[20] However, for these programs to be particularly effective, students need the opportunity to reflect on how different applications of engineering fit for who they are and what they want to achieve. Helping students address this "fit" question is crucial in supporting student development and cannot be done by the engineering faculty alone; it calls for partnerships across the university, with professional societies, and with industry (Sheppard et al. 2009).

Second, it suggests that it is important to increase public understanding that engineering is not a monolithic enterprise in the academy or in the workforce. Engineering work generally involves solving technical-based problems, often on distributed and/or multidisciplinary teams, but the particulars can greatly affect what the "lived" engineering experience is. While this makes it more complicated to explain what engineering is to a young person considering which academic and professional pathway might be right for them, it also provides the opportunity to illustrate how an engineering pathway can be customized for an individual's particular interests, strengths, and goals, and to change perceptions of engineering as "one-size-fits-all," to engineering as a professional pathway that opens up many options.[21]

20. The expansion of hands-on design courses, starting in the first year of college, is one step in this direction (Sheppard et al. 2009), as are a greater variety of extracurricular experiences (e.g., Engineers without Borders, Solar Car projects, Design for America, etc.) and mentoring programs (e.g., MentorNet).

21. This perception work needs to start in middle and high school, where academic choices are being made that affect later options, and builds on calls to "change the conversation" about what engineering encompasses (Committee on Public Understanding of Engineering Messages, National Academy of Engineering 2008). In presenting engineering as an expansive realm of options rather than a narrow realm of technical problem solving, this also may help to recon-

Third, our findings indicate that it may be appropriate for engineering programs to frame their thinking around "our program graduates individuals capable of engineering thinking," rather than "our program graduates engineers." This might liberate programs from needing to cover a long list of "essential knowledge domains," and help them to focus more on the ways of approaching and identifying problems that are unique to the practices of engineering and applicable in a variety of fields.

Fourth, our analysis directs attention at the need for more detailed investigation of linkages between "persistence in the major" and "persistence in the profession," including longitudinal studies of engineering graduates' pathways into nonengineering as well as engineering fields (e.g., via master's- or PhD-level training in business, biological sciences, and information sciences). In terms of attracting a wider group of students into engineering, examining how lower SES students conceive of engineering in the environments in which they tend to matriculate seems critical to understanding the role of socioeconomic characteristics in the development of the engineering workforce. Further research also needs to tease out the interaction between students' financial motivations and perceptions of the labor market, and more disaggregated labor market characteristics than our state-level BLS data.

Finally, we need to know more about what happens in the actual span of time between engineering majors' final years in their programs and their entry into the workforce, which can help schools of engineering better prepare their graduates for what is ahead. Our examination of outcomes in a multilevel framework opens the door for further work on the impact and interactions of personal and institutional factors on engineering career pathways.

Appendix A

Participating APPLES Institutions by Carnegie Classification

The 2000 Carnegie Classification (Carnegie Foundation for the Advancement of Teaching 2001) was used as the basis for the APPLES stratified institutional sampling plan (see Donaldson et al. 2008). The updated 2010 Carnegie Classification categories are presented for comparison.

---

ceptualize engineering as a field with multiple points of entry, rather than one lockstep path that begins at a very young age. Institutions with different resource supports and constraints presumably have much to learn from one another about supporting dynamic pathways and diverse students. Cross-institution conversation can reduce the start-up costs of new initiatives and contribute to more equitable opportunity structures for aspiring engineers.

Table 2A.1 **Updated 2010 Carnegie Classification categories**

Institution	2000 Carnegie Classification	Basic Carnegie Classification (2010)
Alabama A&M University	Doctoral/research intensive	Master's L: Master's colleges and universities (larger programs)
Arizona State University	Doctoral/research extensive	RU/VH: Research universities (very high research activity)
CUNY New York City College of Technology	Master's colleges and universities I	Bac./assoc: Baccalaureate/associate colleges
Columbia University	Doctoral/research extensive	RU/VH: Research universities (very high research activity)
Florida Atlantic University	Doctoral/research intensive	RU/H: Research universities (high research activity)
Georgia Institute of Technology	Doctoral/research extensive	RU/VH: Research universities (very high research activity)
Harvey Mudd College	Baccalaureate colleges— Liberal arts	Bac./A&S: Baccalaureate colleges— Arts & sciences
Kettering University	Specialized institutions— Schools of engineering and technology	Master's M: Master's colleges and universities (medium programs)
Massachusetts Institute of Technology	Doctoral/research extensive	RU/VH: Research universities (very high research activity)
Montana Tech of the University of Montana	Specialized institutions— Schools of engineering and technology	Bac./diverse: Baccalaureate colleges— Diverse fields
North Carolina A&T State University	Master's colleges and universities I	DRU: Doctoral/research universities
Northwestern University	Doctoral/research extensive	RU/VH: Research universities (very high research activity)
Oklahoma Christian University	Baccalaureate colleges— General	Master's M: Master's colleges and universities (medium programs)
Franklin W. Olin College of Engineering	Not classified	Spec./Eng.: Special focus institutions—Schools of engineering
Portland State University	Doctoral/research intensive	RU/H: Research universities (high research activity)
Purdue University	Doctoral/research extensive	RU/VH: Research universities (very high research activity)
San Jose State University	Master's colleges and universities I	Master's L: Master's colleges and universities (larger programs)
Smith College	Baccalaureate colleges— Liberal arts	Bac./A&S: Baccalaureate colleges— Arts & sciences
University of Minnesota– Twin Cities	Doctoral/research extensive	RU/VH: Research universities (very high research activity)
University of Texas at El Paso	Doctoral/research intensive	RU/H: Research universities (high research activity)
West Virginia University Institute of Technology	Baccalaureate colleges— General	Bac./diverse: Baccalaureate colleges— Diverse fields

Appendix B

Composite Measures Used in Multivariate Models

(Question numbers refer to those on the APPLES instrument)

Motivation: Financial ($\alpha = .81$)

Q9b. Reason for pursuing engineering studies: Engineers make more money than most other professionals[a]

Q9e. Reason for pursuing engineering studies: Engineers are well paid[a]

Q9g. Reason for pursuing engineering studies: An engineering degree will guarantee me a job when I graduate[a]

Motivation: Intrinsic Psychological ($\alpha = .75$)

Q9k. Reason for pursuing engineering studies: I feel good when I am doing engineering[a]

Q9m. Reason for pursuing engineering studies: I think engineering is fun[a]

Q9o. Reason for pursuing engineering studies: I think engineering is interesting[a]

Confidence in Professional and Interpersonal Skills ($\alpha = .82$)

Q11a. Self-rating compared to your classmates: Self-confidence (social)[b]

Q11b. Self-rating compared to your classmates: Leadership ability[b]

Q11c. Self-rating compared to your classmates: Public-speaking ability[b]

Q11f. Self-rating compared to your classmates: Communication skills[b]

Q11h. Self-rating compared to your classmates: Business ability[b]

Q11i. Self-rating compared to your classmates: Ability to perform in teams[b]

Academic Involvement—Engineering-Related Courses ($\alpha = .71$)

Q16a. Frequency during current school year: Came late to engineering class (reverse-coded)[c]

Q16b. Frequency during current school year: Skipped engineering class (reverse-coded)[c]

Q16c. Frequency during current school year: Turned in engineering assignments that did not reflect your best work (reverse-coded)[c]

Q16d. Frequency during current school year: Turned in engineering assignments late (reverse-coded)[c]

[a]Four-item scale: 0 = not a reason, 1 = minimal reason, 2 = moderate reason, and 3 = major reason.

[b]Five-point scale: 0 = lowest 10 percent, 1 = below average, 2 = average, 3 = above average, and 4 = highest 10 percent.

[c]Four-point scale: 0 = never, 1 = rarely, 2 = occasionally, and 3 = frequently. Reverse-coded for computation.

Computing the Multi-item Variable Scores

To compute each score, item scores were summed; the scale was then normalized and multiplied by 100 for reporting.

References

Adelman, Clifford. 1998. *Women and Men of the Engineering Path: A Model for Analyses of Undergraduate Careers.* Washington, D.C.: U.S. Department of Education and the National Institute for Science Education.

Amelink, Catherine T., and Elizabeth G. Creamer. 2010. "Gender Differences in Elements of the Undergraduate Experience that Influence Satisfaction with the Engineering Major and the Intent to Pursue Engineering as a Career." *Journal of Engineering Education* 99 (1): 81–92.

American Society for Engineering Education (ASEE). 2011. "Engineering and Engineering Technology College Profiles for 2006." Accessed Sept. 1, 2012. http://profiles.asee.org/.

———. 2015. "Engineering and Engineering Technology College Profiles." Accessed Mar. 6, 2015. http://profiles.asee.org/.

Association of American Medical Colleges. 2011. Table 27: Total Graduates by U.S. Medical School and Sex, 2007–2011. Washington, D.C. Accessed Sept. 4, 2012. https://www.aamc.org/download/145438/data/table27-grad-0711.pdf.

Astin, Alexander W. 1993. *What Matters in College: Four Critical Years Revisited.* San Francisco, CA: Jossey Bass.

Astin, Alexander W., and Leticia Oseguera. 2004. "The Declining 'Equity' of American Higher Education." *Review of Higher Education* 27 (3): 321–41.

Atman, Cynthia J., Sheri D. Sheppard, Jennifer Turns, Robin S. Adams, Lorraine N. Fleming, Reed Stevens, Ruth A. Streveler, Karl A. Smith, Ronald L. Miller, Larry J. Leifer, Ken Yasuhara, and Dennis Lund. 2010. *Enabling Engineering Student Success: The Final Report for the Center for the Advancement of Engineering Education.* San Rafael, CA: Morgan & Claypool Publishers.

Besterfield-Sacre, Mary, Cindy J. Atman, and Larry J. Shuman. 1997. "Characteristics of Freshmen Engineering Students: Models for Determining Student Attrition in Engineering." *Journal of Engineering Education* 86 (2): 139–49.

Besterfield-Sacre, Mary E., Magaly Moreno, Larry J. Shuman, and Cynthia J. Atman. 2001. "Gender and Ethnicity Differences in Freshmen Engineering Student Attitudes: A Cross-Institutional Study." *Journal of Engineering Education* 90 (4): 477–89.

Bradburn, Ellen M., Stephanie Nevill, Emily Forrest Cataldi, and Kristin Perry. 2006. *Where Are They Now? A Description of 1992–93 Bachelor's Degree Recipients 10 Years Later.* NCES 2007-159. Washington, D.C.: U.S. Department of Education, National Center for Education Statistics.

Brainard, Suzanne G., and Linda Carlin. 1998. "A Six-Year Longitudinal Study of Undergraduate Women in Engineering and Science." *Journal of Engineering Education* 87 (4): 369–75.

Brawner, Catherine E., Michelle M. Camacho, Russell A. Long, Susan M. Lord, Matthew W. Ohland, and Mara H. Wasburn. 2009. "Work in Progress—The Effect of Engineering Matriculation Status on Major Selection." In *Proceedings of the IEEE/ASEE Frontiers in Education Conference*, San Antonio, TX, Oct. 18–21.

Brawner, Catherine E., Michelle M. Camacho, Susan M. Lord, Russell A. Long, and Matthew W. Ohland. 2012. "Women in Industrial Engineering: Stereotypes, Persistence, and Perspectives." *Journal of Engineering Education* 101 (2): 288–318.

Brunhaver, Samantha, Shannon Gilmartin, Helen L. Chen, Michelle Grau, Michelle Warner, Holly Matusovich, Katherine Winters, Cheryl Carrico, and Sheri Sheppard. 2013. "Differences in the Work Characteristics and Experiences of Early Career Engineering Graduates." Paper presented at the American Educational Research Association Annual Meeting, San Francisco, CA, Apr. 27–May 1.

Brunhaver, Samantha, Shannon Gilmartin, Helen L. Chen, and Sheri Sheppard. 2012. "Factors Associated with the Current Occupations of Early Career Engineering Graduates." Paper presented at the Association for the Study of Higher Education Annual Conference, Las Vegas, NV, Nov. 14–17.

Bureau of Labor Statistics (BLS). 2011. "National Cross-Industry Estimates, May 2007." Accessed Sept. 1, 2012. http://www.bls.gov/oes/2007/may/oes_dl.htm#2007.

Carnegie Foundation for the Advancement of Teaching. 2001. *The Carnegie Classifications of Institutions of Higher Education*, 2000 ed. Stanford, CA: Carnegie Foundation for the Advancement of Teaching.

Carnevale, Anthony P., Stephen J. Rose, and Ban Cheah. 2011. *The College Payoff: Education, Occupations, Lifetime Earnings*. Washington, D.C.: Georgetown University Center on Education and the Workforce.

Carnevale, Anthony P., Jeff Strohl, and Michelle Melton. 2011. *What's It Worth? The Economic Value of College Majors*. Washington, D.C.: Georgetown University Center on Education and the Workforce.

Chen, Helen L., Krista Donaldson, Özgür Eris, Debbie Chachra, Gary Lichtenstein, Sheri D. Sheppard, and George Toye. 2008. "From PIE to APPLES: The Evolution of a Survey Instrument to Explore Engineering Student Pathways." In *Proceedings of the American Society for Engineering Education Annual Conference*, Pittsburgh, PA, June 22–25.

Choy, Susan P., Ellen M. Bradburn, and C. Dennis Carroll. 2008. *Ten Years after College: Comparing the Employment Experiences of 1992–93 Bachelor's Degree Recipients with Academic and Career-Oriented Majors*. NCES 2008-155. Washington, D.C.: U.S. Department of Education, National Center for Education Statistics.

Clark, Mia, Sheri D. Sheppard, Cynthia J. Atman, Lorraine Fleming, Ronald Miller, Reed Stevens, Ruth Streveler, and Karl Smith. 2008. "Academic Pathways Study: Processes and Realities." In *Proceedings of the American Society for Engineering Education Annual Conference*, Pittsburgh, PA, June 22–25.

Colbeck, Carol L., Susan E. Campbell, and Stefani A. Bjorklund. 2000. "Grouping in the Dark: What College Students Learn from Group Projects." *Journal of Higher Education* 71 (1): 60–83.

Committee on Public Understanding of Engineering Messages, National Academy of Engineering. 2008. *Changing the Conversation: Messages for Improving Public Understanding of Engineering*. Washington, D.C.: National Academies Press.

Donaldson, Krista, Helen L. Chen, George Toye, Mia Clark, and Sheri D. Sheppard. 2008. "Scaling Up: Taking the Academic Pathways of People Learning Engineering Survey (APPLES) National." In *Proceedings of the IEEE/ASEE Frontiers in Education Conference*, Saratoga Springs, NY, Oct. 22–25.

Donaldson, Krista, Helen L. Chen, George Toye, and Sheri Sheppard. 2007. "Targeting Undergraduate Students for Surveys: Lessons from the Academic Pathways of People Learning Engineering Survey (APPLES)." In *Proceedings of the 37th Annual Frontiers in Education Conference*, Milwaukee, WI, Oct. 10–13.

Eris, Özgür, Debbie Chachra, Helen L. Chen, Camelia Rosca, Larry Ludlow, Sheri D. Sheppard, and Krista Donaldson. 2007. "A Preliminary Analysis of Correlates of Engineering Persistence: Results from a Longitudinal Study." In *Proceedings of the American Society for Engineering Education Annual Conference*, Honolulu, HI, June 24–27.

Eris, Özgür, Debbie Chachra, Helen L. Chen, Sheri D. Sheppard, Larry Ludlow, Camelia Rosca, Tori Bailey, and George Toye. 2010. "Outcomes of a Longitudinal Administration of the Persistence in Engineering Survey." *Journal of Engineering Education* 99 (4): 371–95.

Eris, Özgür, Helen L. Chen, Tori Bailey, Kimarie Engerman, Heidi Loshbaugh, Ashley Griffin, Gary Lichtenstein, and Angela Cole. 2005. "Development of the Persistence in Engineering (PIE) Survey Instrument." In *Proceedings of the American Society for Engineering Education Annual Conference*, Portland, OR, June 12–15.

Forrest Cataldi, Emily, Caitlin Green, Robin Henke, Terry Lew, Jenny Woo, Bryan Shepherd, Peter Siegel, and Ted Socha. 2011. "2008–09 Baccalaureate and Beyond Longitudinal Study (B&B: 08/09): First Look." Washington, D.C.: U.S. Department of Education, National Center for Education Statistics.

Frehill, Lisa. 2007a. "The Society of Women Engineers National Survey about Engineering: Are Women More or Less Likely Than Men to Be Retained in Engineering after College?" *SWE Magazine* 53 (4): 22–25.

————. 2007b. "What do Women do with Engineering Degrees?" In *Proceedings of the 2007 Women in Engineering Programs and Advocates Network (WEPANP) Conference*, Lake Buena Vista, FL, June 10–13.

Garrison, Lari, Reed Stevens, Portia Sabin, and Andrew Jocuns. 2007. "Cultural Models of the Admission Process in Engineering: Views on the Role of Gender, Women in Engineering." In *Proceedings of the American Society for Engineering Education Annual Conference*, Honolulu, HI, June 24–27.

Herrera, Felisha A., and Sylvia Hurtado. 2011. "Maintaining Initial Interests: Developing Science, Technology, Engineering, and Mathematics (STEM) Career Aspirations among Underrepresented Racial Minority Students." Paper presented at the annual meeting of the American Educational Research Association, New Orleans, LA, Apr. 8–12.

Huang, Gary, Nebiyu Taddese, and Elizabeth Walter. 2000. "Entry and Persistence of Women and Minorities in College Science and Engineering Education, NCES 2000-601." Washington, D.C.: U.S. Department of Education. National Center for Education Statistics.

Lichtenstein, Gary, Heidi G. Loshbaugh, Brittany Claar, Helen L. Chen, Kristyn Jackson, and Sheri D. Sheppard. 2009. "An Engineering Major Does Not (Necessarily) an Engineer Make: Career Decision-Making among Undergraduate Engineering Majors." *Journal of Engineering Education* 98 (3): 227–34.

Lord, Susan M., Catherine E. Brawner, Michelle M. Camacho, Richard A. Layton, Russell A. Long, Matthew W. Ohland, and Mara H. Wasburn. 2008. "Work in Progress: A Study to Investigate the Effect of Climate and Pedagogy on Persistence of Women in Undergraduate Engineering Programs." In *Proceedings of the IEEE/ASEE Frontiers in Education Conference*, Saratoga Springs, NY, Oct. 22–25.

Lord, Susan M., Michelle M. Camacho, Richard A. Layton, Russell A. Long, Matthew W. Ohland, and Mara H. Wasburn. 2009. "Who's Persisting in Engineering? A Comparative Analysis of Female and Male Asian, Black, Hispanic, Native American and White Students." *Journal of Women and Minorities in Science and Engineering* 15 (2): 167–90.

Lowell, B. Lindsay, Hal Salzman, Hamutal Bernstein, and Everett Henderson. 2009. "Steady as She Goes? Three Generations of Students through the Science and

Engineering Pipeline." Paper presented at the Annual Meetings of the Association for Public Policy Analysis and Management, Washington, D.C., Nov. 7.

Margolis, James, and Daria Kotys-Schwartz. 2009. "The Post-Graduation Attrition of Engineering Students: An Exploratory Study on Influential Career Choice Factors." In *Proceedings of the American Society of Mechanical Engineers International Mechanical Engineering Congress and Exposition*, Lake Buena Vista, FL, Nov. 13–19.

National Association for Law Placement. 2012a. "Class of 2011 National Summary Report." Washington, D.C. Accessed Sept. 4, 2012. http://www.nalp.org/uploads/NatlSummChart_Classof2011.pdf.

———. 2012b. "Employment for the Class of 2011—Selected Findings." Washington, D.C. Accessed Sept. 4, 2012. http://www.nalp.org/uploads/Classof2011SelectedFindings.pdf.

National Center for Education Statistics. 2011. "Integrated Postsecondary Education Data System, 2006." Accessed Sept. 1, 2012. http://nces.ed.gov/ipeds/datacenter/Default.aspx.

National Resident Matching Program and Association of American Medical Colleges. 2011. Chart 1: Active Applicants in the 2011 Main Residency Match by Applicant Type (page 2). Charting Outcomes in the Match, 2011. Washington, D.C. Accessed Sept. 4, 2012. http://www.nrmp.org/data/chartingoutcomes2011.pdf.

National Science Foundation (NSF), Division of Science Resources Statistics. 2011a. *Women, Minorities, and Persons with Disabilities in Science and Engineering: 2011*. Special Report no. NSF 11-309, Arlington, VA. Accessed Sept. 8, 2012. http://www.nsf.gov/statistics/wmpd/.

National Science Foundation, Division of Science Resources Statistics. 2011b. Scientists and Engineers Statistical Data System (SESTAT), 2006 National Survey of Recent College Graduates. Arlington, VA. Accessed Aug. 10, 2012. https://www.nsf.gov/statistics/sestat.

National Science Foundation, Division of Science Resources Statistics. 2015. *Women, Minorities, and Persons with Disabilities in Science and Engineering*. Arlington, VA. Accessed Mar. 6, 2015. http://www.nsf.gov/statistics/wmpd/.

National Science Foundation, National Center for Science and Engineering Statistics. 2012. SESTAT PUBLIC 1993 Integrated Survey Data. Unpublished tabulation. http://www.nsf.gov/statistics/sestat.

Ohland, Matthew W., Catherine E. Brawner, Michelle M. Camacho, Richard A. Layton, Russell A. Long, Susan M. Lord, and Mara H. Wasburn. 2011. "Race, Gender, and Measures of Success in Engineering Education." *Journal of Engineering Education* 100 (2): 225–52.

Ohland, Matthew W., Michelle M. Camacho, Richard A. Layton, Susan M. Lord, and Mara H. Wasburn. 2009. "How We Measure Success Makes a Difference: Eight-Semester Persistence and Graduation Rates for Female and Male Engineering Students." In *Proceedings of the American Society for Engineering Education Annual Conference*, Austin, TX, June 14–17.

Ohland, Matthew W., Sheri D. Sheppard, Gary Lichtenstein, Özgür Eris, Debbie Chachra, and Richard A. Layton. 2008. "Persistence, Engagement, and Migration in Engineering." *Journal of Engineering Education* 97 (3): 259–78.

Orr, Marisa, Zahra Hazari, Philip Sadler, and Gerhard Sonnert. 2009. "Career Motivations of Freshman Engineering and Non-Engineering Students: A Gender Study." In *Proceedings of the American Society for Engineering Education Annual Conference*, Austin, TX, June 14–17.

Perna, Laura W. 2003. "The Private Benefits of Higher Education: An Examination of the Earnings Premium." *Research in Higher Education* 44 (4): 451–72.

Petrides, Lisa A. 1996. "A Study of the Gendered Construction of the Engineering Academic Context in Graduate School." PhD diss., Stanford University.

Raudenbush, Stephen W., and Anthony S. Bryk. 2002. *Hierarchical Linear Models: Applications and Data Analysis Methods*, 2nd ed. Thousand Oaks, CA: Sage.

Reardon, Sean. F., Rachel Baker, and Daniel Klasik. 2012. "Race, Income, and Enrollment Patterns in Highly Selective Colleges, 1982–2004." Stanford, CA. Accessed Sept. 4, 2012. http://cepa.stanford.edu/content/race-income-and -enrollment-patterns-highly-selective-colleges1982-2004.

Reese, Carol. 2003. "Employment History Survey of ASCE's Younger Members." *Leadership and Management in Engineering* 3 (1): 33–53.

Regets, Mark C. 2006. "What Do People Do after Earning a Science and Engineering Bachelor's Degree?" Info Brief no. 06-234, Washington, D.C., National Science Foundation.

Ro, Hyun Kyoung. 2011. "An Investigation of Engineering Students' Post-Graduation Plans inside or outside of Engineering." PhD diss., Pennsylvania State University.

Robst, John. 2007. "Education and Job Match: The Relatedness of College Major and Work." *Economics of Education Review* 26 (4): 397–407.

Rouse, Cecilia Elena. 2004. "Low-Income Students and College Attendance: An Exploration of Income Expectations." *Social Science Quarterly* 85 (5): 1299–317.

Salzman, Hal, and Leonard Lynn. 2010. "Engineering and Engineering Skills: What's Really Needed for Global Competitiveness." Paper presented at the Association for Public Policy Analysis and Management Annual Meetings, Boston, MA, Nov. 4.

Seymour, Elaine, and Nancy M. Hewitt. 1997. *Talking about Leaving: Why Undergraduates Leave the Sciences*. Boulder, CO: Westview Press.

Sheppard, Sheri D., Anthony L. Antonio, Samantha R. Brunhaver, and Shannon K. Gilmartin. 2014. "Studying the Career Pathways of Engineers: An Illustration with Two Datasets." In *Cambridge Handbook of Engineering Education Research*, edited by Aditya Johri and Barbara Olds, 283–309. New York: Cambridge University Press.

Sheppard, Sheri D., Shannon Gilmartin, Helen L. Chen, Krista Donaldson, Gary Lichtenstein, Özgür Eris, Micah Lande, and George Toye. 2010. *Exploring the Engineering Student Experience: Findings from the Academic Pathways of People Learning Engineering Survey (APPLES)* (TR-10-01). Seattle, WA: Center for the Advancement for Engineering Education.

Sheppard, Sheri D., Kelly Macatangay, Anne Colby, and William M. Sullivan. 2009. *Educating Engineers: Designing for the Future of the Field*. Stanford, CA: Carnegie Foundation for the Advancement of Teaching.

Titus, Marvin A. 2006. "Understanding College Degree Completion of Students with Low Socioeconomic Status: The Influence of the Institutional Financial Context." *Research in Higher Education* 47 (4): 371–98.

3

Engineering Educational Opportunity
Impacts of 1970s and 1980s Policies to Increase the Share of Black College Graduates with a Major in Engineering or Computer Science

Catherine J. Weinberger

During the 1970s and 1980s, in the context of a national conversation about racial inequality, a constellation of overlapping policy efforts aimed to increase the number of black college graduates with engineering or computer science majors. These efforts can be roughly organized into two categories. The first was inspired by business leaders concerned about the absence of black Americans among top executives, and cognizant that many of the most influential industrial leaders had engineering training. This effort involved representatives of industry, private foundations, and educators from campuses across the United States, including the six engineering programs on historically black campuses, which were graduating the majority of black engineers at the time. The second came out of the desire to improve the future career prospects of graduates within a broader coalition of historically black colleges and universities (HBCUs),[1] including those with no engineering program. This second wave effort strove to expand opportunities for both men and women to study engineering, and led to new computer science programs at many HBCU campuses.

Catherine J. Weinberger is an independent scholar affiliated with the Institute for Social, Behavioral and Economic Research (ISBER), the Broom Center for Demography, and the department of economics at the University of California, Santa Barbara.

This material is based upon work supported by the National Science Foundation under Grant no. 0830362. The use of NSF data does not imply NSF endorsement of the research methods or conclusions. Opinions, findings, and conclusions or recommendations expressed here are those of the author and do not necessarily reflect the views of the National Science Foundation. I thank Mark Woolley for exceptional research assistance over an extended period of time. Orel Marzini and Anand Shukla also assisted with this project. For acknowledgments, sources of research support, and disclosure of the author's material financial relationships, if any, please see http://www.nber.org/chapters/c12685.ack.

1. For statistical purposes, HBCU is defined as an institution where the majority of students were black immediately before passage of the 1964 Civil Rights Act.

This chapter shows that these campaigns succeeded in raising the proportion of college-educated black men and women entering engineering and computer science fields. Initially, the change was particularly rapid among persons born near the six historically black engineering programs. These engineering programs were able to increase the number of black graduates quickly, relative to most other U.S. institutions. While there remain black-white gaps in engineering despite changes at engineering campuses across the country, the strengthening of the HBCUs decades ago contributed and continues to contribute to limiting this gap. The chapter also shows that the second wave effort to introduce computer science courses at HBCU campuses was so successful that black college graduates became more likely than the U.S. average to hold a computer science degree. Because there was little change in the relative propensity to major in math or other science fields, growing engineering and computer science participation led to a net increase in the representation of black college graduates among science, technology, engineering, and mathematics (STEM) majors.[2]

The chapter is divided into four sections. The first part describes historical features of the higher education infrastructure and details of national and local efforts to change. The second part presents my analysis of the impacts of these efforts on the number of black engineering and computer science graduates in each year, based on data describing the number and type of degrees conferred by each institution of higher education over the interval 1968–2011, collected by the U.S. Department of Education and the Engineering Manpower Commission. To establish a link between the geography of institution-level changes and the impacts on college students from different states, I draw on information about college major plus year and state of birth from the nationally representative American Community Survey (ACS) 2009–2013. The third part examines the occupations and earnings of college graduates in current ACS data to learn how changes in the number of black Americans with engineering and computer science education translated into current labor market outcomes. The fourth part concludes.

3.1   Historical Context of the Higher Education Infrastructure

Prior to the civil rights era, the most prevalent job opportunity for black college graduates was to teach in a segregated school, and educational offerings at HBCU campuses tended to reflect this reality (U.S. Office of Education 1942/1943; Weinberg 1977; Pruitt 1987). The separate-but-equal doctrine associated with the 1896 Supreme Court decision in *Plessy v. Ferguson* preserved racial segregation, which eventually led to the 1954 *Brown v. Board*

---

2. STEM is shorthand for science, technology, engineering, and mathematics. Here it includes biological and physical sciences, computer and information sciences, engineering, mathematics and statistics.

of Education ruling that separate is inherently unequal. The Civil Rights Act of 1964 provided further promise of access to education for black students, but systems of higher education remained separate and unequal, particularly in the South where HBCU programs remained poorly funded. No specific legal rulings addressed equitable funding for black college campuses or desegregation of undergraduate programs until a string of decisions in the far less familiar *Adams* case that spanned 1972–1983 eventually required nineteen states to submit plans of remedial action to the federal government (Pruitt 1987). The wording of this requirement recognized the unique role played by HBCU campuses as part of the U.S. educational infrastructure (Pruitt 1987).

The higher-education landscape in the United States was largely in place by the early nineteenth century (Goldin and Katz 1999). While the *Adams* case was being argued, black students in most southern states had limited access to engineering education. During the 1960s and 1970s, only six of the HBCU campuses housed an accredited engineering program. These were four public land-grant universities, which enrolled primarily in-state students: Prairie View A&M in Texas, Southern University and A&M College in Louisiana, North Carolina A&T State University, and Tennessee State (formerly Agricultural and Industrial) University,[3] plus Howard University in Washington, D.C., and Tuskegee in Alabama, HBCUs with a somewhat wider geographic pull.[4] No federal statistics on the number of black engineering graduates were collected until 1968, but it was estimated that at least one-half of black engineers in the United States were trained at one of these six HBCU campuses (HBCU6) during the 1960s and earlier (Pierre 1972).[5]

On many HBCU campuses, a commitment to support the intellectual development of students regardless of prior academic preparation is viewed as a legacy of the historic mission to teach emancipated slaves how to read. During the 1970s and 1980s a group of HBCU campuses organized to expand educational opportunities in engineering, computer science, and other technical fields, "to prepare their students for expanded career choices" (Trent and Hill 1994). Contemporaneous observers describe an encouraging pedagogical environment that exemplifies the heart of what education can be: "They take students who may not have been well prepared in high school for careers in the 'hard sciences' and graduate them with degrees in science and engineering" (Trent and Hill 1994). The success of

3. For historical background on the establishment of land grant universities, see Weinberg (1977). It is notable that the majority of the HBCU engineering campuses began as agricultural and technical, mechanical, or industrial campuses.
4. All six campuses offered engineering degrees by 1960; one has been training engineers since 1912.
5. The exact share is impossible to determine because educational statistics on new degrees conferred by race were not collected before 1968. My estimate of two-thirds, given later in this chapter, is based on three different bodies of data. It exceeds the estimate of at least one-half mentioned at the time.

both early and current educational programs at HBCU campuses has been well documented. While college-completion rates tend to increase with the selectivity of an institution, HBCUs have graduation rates far higher than comparably selective colleges (Kane 1998). Other superior student outcomes include shorter time to graduation, and encouragement and academic support to pursue more challenging and remunerative college majors compared to students with observably similar characteristics at other colleges (Trent and Hill 1994; Ehrenberg and Rothstein 1994; Nettles 1988; Kane 1998; Sibulkin and Butler 2005; Weinberger and Joy 2007). The supportive HBCU environment is described in interviews with students who took courses at both an HBCU campus and another institution of higher education (Fries-Britt, Burt, and Franklin 2012), and also by the former dean of the Howard University School of Engineering (Pierre 1972). As summarized by Slaughter (2009), historically black schools of engineering provide an environment "in which success is encouraged, supported, and expected."

In contrast, the first black students who enrolled at Georgia Tech in the early 1960s recall an unsupportive educational environment.[6] Relying on primary historical documents and interviews, Bix (2013) notes that these students faced social barriers with profound implications for their academic development including difficulty finding lab partners, exclusion from fraternities and their valuable files of prior years' exams, a "ring of empty seats around me in class," and threats of physical violence that rendered the library inaccessible (Bix 2013). It is telling that the first black student to persist to graduation had transferred from, and continued to find social support within, a nearby HBCU campus. Landis (2005, 6)—who visited dozens of engineering campuses throughout the 1970s and 1980s—reported that, despite the best of institutional intentions, "At university after university, minority engineering students have told me that white students won't form laboratory groups with them, act surprised when they do well on tests, and intentionally leave the seats next to them vacant." Today's STEM undergraduates report similar experiences: "there was no one willing to be my lab partner. . . . They don't think I'm capable enough or know the material. What it means is I study alone" (Abcarian 2017). Landis (2005) argues that fostering strong social support and a collaborative learning environment is key to successfully retaining capable black engineering students. The successful programs he developed are based on his own experiences as the member of an academically oriented fraternity while he was an engineering student at MIT (Landis 2005). Highlighting the importance of social support, Treisman (1992) observed in the 1970s that academically talented black students at UC Berkeley did not learn as much calculus as ethnically Chinese classmates because of their tendency to study alone rather than with groups of friends. Successful programs of social and

---

6. Since that time, Georgia Tech has graduated thousands of black engineers.

academic engagement and support, based on models inspired by these and other educators, are thriving on many engineering campuses today.

3.1.1   The 1970s Intervention

The beginning of the nationwide effort to increase opportunities for black students to enter engineering is attributed to a speech "Needed: Revolutionary Approaches Leading to Minority Management Development," by General Electric (GE) executive J. Stanford Smith in 1972.[7] The speech was given to a group of corporate executives and forty-four engineering school deans at the GE Management Development Center in Crotonville, NY, during a five-day conference on "Strategic Considerations in Engineering Education" (Lusterman 1979; Blackwell 1981). It emphasized that the majority of top leaders in industrial management began their careers with engineering study, and accumulated many years of engineering experience before rising to leadership, and made the case that integrating top management twenty years in the future would require immediate action to increase the supply of black engineers (Smith 1973). Lusterman's (1979) account, written shortly afterward, describes how GE analysts, charged with figuring out how to meet hiring goals, determined that there was a bottleneck in the supply chain that should be remedied. Lusterman (1979) reports that the audience was "startled" by Smith's presentation of GE's analysis, and his call for "an undertaking of staggering proportions that requires revolutionary action." A representative of the Alfred P. Sloan Foundation and a science advisor to President Nixon were also in attendance.[8] In concluding, Smith—an industry leader with a family legacy of advocacy for educational opportunity—called for a national effort to rectify inequities in engineering education, with cooperation of the business and education communities, foundations, and professional societies (Lusterman 1979; Blackwell 1981).[9] Senator Humphrey felt the speech was so important that he summarized its main points in a brief oration addressed to President Nixon, and had the text of the entire speech (with revised title) entered into the U.S. Congressional Record, following his own remarks (Humphrey 1973). The Crotonville speech continues to be referenced by those working to rectify inequities in access to engineering education.

Although many inspiring speeches are delivered to little long-term effect,

7. Soon after the passage of the 1964 Civil Rights Act, there were some modest efforts to increase the number of black engineering graduates. Some companies seeking additional engineers donated funds to improve capacity and quality at the HBCU6 campuses, and to establish "dual degree" programs between other HBCU campuses and engineering schools prepared to accept transfer students after the first two or three years of study (Pierre 1975; Lusterman 1979). But these efforts were insufficient to change the majority of educational institutions.

8. Lucius P. Gregg completed a master's degree at MIT after graduating with distinction from the Naval Academy in 1955 (Williams 1999; Schneller 2008).

9. J. Stanford Smith's mother served on the advisory board of the Mary McLeod Bethune School for Negro Girls; his daughter was influenced by discussions about educational equity around the family dinner table, and continues to advocate for educational opportunity as a scholar and professor (Witherell 2009; College of Lewis and Clark 2012).

the Crotonville speech began conversations that affected private and public policy. Discussions between GE representatives and the educators in attendance led to deeper understanding of broad challenges, including the need to improve in K–12 education (Pierre 2013, 2015). As the scope of the necessary effort became clear, Lindon Saline, director of the GE Management Development Institute, enlisted the participation of Percy Pierre, who attended the Crotonville conference as the Dean of Engineering at Howard University. After approaching several organizations without success, this pair persuaded the Commission on Education of the National Academy of Engineering to host a Symposium on Minorities in Engineering, and pulled in financial sponsors including the Sloan Foundation (NAE 1973; Slaughter 2009; Pierre 2013, 2015). At this symposium, held four months after Smith's Crotonville speech, consensus was reached on building a national organization and extending the institutional support of the National Academy of Engineering toward this effort, with the cooperation of GE and other corporations (Slaughter 2009; Pierre 2013, 2015). The Academy created the National Advisory Council for Minorities in Engineering (NACME), with GE Chairman Reginald Jones as its first chair (Pierre 2013).[10] Soon afterward the Sloan Foundation committed 20 percent of its resources over five to seven years ($12–$15 million) toward this national effort, and asked Percy Pierre to serve as program officer to oversee Sloan's multimillion-dollar investment, which he agreed to do half time while continuing as dean of engineering at Howard University (Lusterman 1979; Blackwell 1981; Pierre 1975, 2012, 2015). Pierre (1975) later reported that the Sloan Foundation was interested in funding such an effort several years earlier, but was waiting for a sign that they would not have to work alone. J. Stanford Smith's speech and the rapid organizational energy that followed soon afterward provided the signals that the Sloan Foundation had been waiting for (Pierre 1975; Pierre 2013).

Over the next year, the newly formed Planning Commission for Expanding Minority Opportunities in Engineering met regularly under the leadership of Professor Louis Padulo of Stanford University, who had previously established a dual degree program between Georgia Tech and nearby HBCU campuses.[11] This group—seventeen representatives of industry, academia, government, and other organizations recruited by the Sloan

10. The National Action Council for Minorities in Engineering that exists in 2016 came from the 1980 merger of the original NACME organization with two closely intertwined organizations, the Minority Engineering Effort (ME3), and the National Fund for Minority Engineering Students (NFMES). While the focus of NACME is on undergraduate education, other organizations formed during this effort focus on graduate level (the National Consortium for Graduate Degrees for Minorities in Engineering, or GEM) and K–12 education (Mathematics, Engineering, Science Achievement, or MESA).

11. According to his web page, Dr. Padulo established the Georgia Tech dual degree engineering program soon after earning a PhD from that institution, while a professor at Morehouse by invitation of Dr. King (Padulo 2015).

Foundation—produced a "Blueprint for Action" that treated the effort as an engineering problem, and recommended approaches to overcome obstacles (PCEMOE 1974; Pierre 1975; Blackwell 1981; Pierre 2013). Suggested mechanisms included funding to expand and improve programs at the HBCU6 institutions, funding to establish additional dual degree partnerships between HBCU and engineering campuses, thousands of undergraduate minority engineering scholarships, expanded support for the transition to graduate school, incentives for individual engineering programs to begin or expand recruitment efforts, and efforts to improve precollege math and science preparation across the nation (PCEMOE 1974; Pierre 1975; Lusterman 1979; Blackwell 1981). The Blueprint report was quickly endorsed by NACME (Pierre 2013). An edited volume of sixteen articles by twenty-one authors gives a sense of the teamwork inspired by the early effort (Saline 1974) that in short order translated recommendations into action as social activists, educators, and engineering corporations reconfigured the opportunity structure in engineering.

The resulting efforts to expand the pool of qualified black engineers were well funded (Blackwell 1981; Lusterman 1979). In addition to the Sloan Foundation's seed money, major engineering employers donated millions of dollars both to the emerging national organizations and to individual university campuses (Lusterman 1979). Donations by potential employers to educational institutions cemented relationships that paid off when it was time to recruit new graduates (Pierre 2012). The donations toward equitable engineering education were also encouraged by the newly formed Equal Employment Opportunity Commission (EEOC) as one of the hallmarks of an EEOC-compliant federal contractor.[12] The large total value of all donations can be surmised by surviving documentation: a 1973 advertisement in *Black Enterprise* states that IBM placed twenty-five full-time engineers in teaching positions at black colleges and donated a large quantity of equipment as well, and a 1986 NACME publication itemizes several million dollars in donations made by a veritable who's who of American industry (IBM 1973; Miranda and Ruiz 1986). Money was distributed both as scholarships to individuals and also in the form of institutional support for minority engineering programs through an "Incentive Grants" program that required institutions to set and meet goals for minority engineering enrollment and graduation (Blackwell 1981, 1987; Miranda and Ruiz 1986). By 1983, nearly half of all U.S. engineering programs were actively recruiting minority students (NACME 1986; Miranda and Ruiz 1986; Blackwell 1987).

Through the 1990s, opportunities for black students to enroll in engi-

12. This is suggested by an observation made by Pierre (1972, 2) that "It is explicitly included in the Executive Order establishing affirmative action that activity by companies to increase the supply of black engineers is part of what the Labor Department must look for. So this is one reason, I think, why we have noticed this interest today."

neering programs continued to grow rapidly. In the southern states, several historically black campuses added accredited engineering programs: Alabama A&M, Florida A&M, University of the District of Columbia, Morgan State (Maryland), Hampton University (Virginia) between 1979 and 1992, joined by Jackson State (Mississippi), South Carolina State, Virginia State, and Norfolk State (Virginia) in the most recent decade.[13] Meanwhile, representation of black engineering students at other institutions continued to expand, so that the addition of new HBCU programs was matched by equally strong trends toward inclusion at engineering programs across the country. While the national effort to train black engineers did not meet the ambitious goals set out by early activists, it produced real and sustained change. Blackwell (1981, 1987), who wrote a book with one chapter devoted to the entry of black professionals into each of ten occupations, concluded that the engineering effort was the most successful among all the professions. More recently, Conrad (2006) has noted substantial variation across geographic regions in the representation of black college graduates in high tech occupations.

3.2   Analysis of the Impacts of the Intervention

3.2.1   Access to Engineering Education

To see how the national effort just described affected the flow of black students into engineering majors nationwide, I examine data on the number of black engineering majors. I begin with broad national counts and then move to more detailed statistics. Current nationally representative ACS 2009–2013 data allow a broad-brush view of the change that occurred because the educational attainment of today's adults reflects the educational opportunities available when each cohort reached college age. Figure 3.1 graphs the share of the U.S. population and of the black population currently holding college degrees, and the share holding college degrees with an engineering major, by birth cohort. Over time, the share of all Americans with engineering degrees fluctuated with labor market conditions but remained close to the nearly 2 percent average, as modeled and described by Freeman (1976b). Over the 1950–2000 interval, increasing shares of successive cohorts of black Americans earned engineering degrees, with fastest growth between the 1970s and 1980s. Even so, across all cohorts, black representation is far below the national average. Among those who reached age twenty-two during the 1950s, only 0.37 percent (1 in 270) trained as engineers. The share

13. Since programs must be operational before they can be evaluated for accreditation, the actual dates of establishment are earlier. Accreditation dates are based on information provided at the website of ABET, Inc., formerly known as the Engineer's Council for Professional Development (ECPD, 1932–1980) and the Accreditation Board for Engineering and Technology (ABET, 1980–2005). (http://www.abet.org/AccredProgramSearch/AccreditationSearch.aspx.)

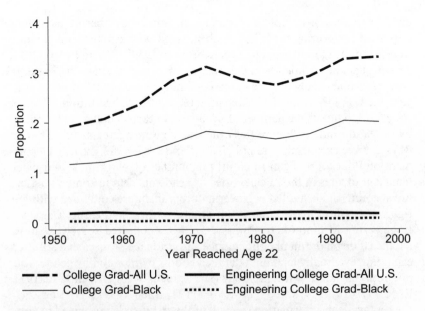

Fig. 3.1   Share of indicated population reporting college degree or college degree with a major in engineering by race and birth cohort (year reached age twenty-two, in five-year intervals)

Sample: American Community Survey 2009–2013, restricted to those born in the United States between 1928 and 1977 (age twenty-two in 1950–1999).

increased over time so that for the cohort that reached age twenty-two in the 1990s it was 0.84 percent, still less than half of the national share but over two times the 1950s ratio—evidence of a diminished racial gap in the propensity to enter engineering. Much of the remaining gap in the 1990s is related to a lower propensity to graduate college. The average propensity to graduate college increased over time, but the relative propensity for the black population remained near 60 percent of U.S. levels throughout the fifty-year span. For the most recent cohort depicted in figure 3.1, 4.15 percent of black college graduates majored in engineering compared to 5.67 percent of U.S. college graduates.[14] This represents three-fourths the national share, up from less than one-third estimated for the earliest cohort depicted. The next question is whether the timing of this shift corresponds to the policy efforts described in the previous section.

To describe changes over time in the number of black engineering grad-

14. To confirm that these estimates based on birth cohort are fairly accurate, I compared the ACS estimates for college graduates who turned twenty-two between the 1950s and the 1970s with comparable estimates for college graduates who actually earned their degrees between the 1950s and the 1970s drawn from the 1993 and 2003 National Surveys of College Graduates. In both cases, I estimate that 2–3 percent of black college graduates majored in engineering during that era, compared to nearly 7 percent of all U.S. graduates.

uates, estimates or complete counts of the annual number of bachelor's degrees in engineering earned by black men and women in the United States were assembled for all academic years between 1959–1960 and 2010–2011. The most detailed statistics on the racial composition of engineering graduates are available only in later years. Beginning in the 1982–1983 academic year, the U.S. Department of Education required each institution of higher education to report the number of graduates in each detailed field of study by race and gender.[15] Before this time, counts were aggregated in different ways. Between academic years 1975–1976 and 1981–1982, the U.S. Department of Education began to count the number of black male and black female graduates in broad categories of academic field in some years, but racial counts at accredited engineering programs were combined with less rigorous engineering technology programs. In these and earlier years, the Engineering Manpower Commission (EMC) collected somewhat more detailed data on the number of engineering graduates from each institution each year, eventually including detailed counts by race and by gender (but not both) beginning with the 1968–1969 academic year (Alden 1970, 1971; EMC 1972, 1977, 1978, 1980, 1981, 1984, 1987, 1988, 1989).[16] For this study, the best available information from all of these sources is combined to paint a complete picture of changes over time.

Before 1968 there are no systematic periodic counts of the number of black engineering graduates, but previous estimates suggest that at least half of all black engineering graduates attended the HBCU6 campuses (Pierre 1972). Due to pervasive segregation at this time, earlier counts of the number of HBCU6 engineering graduates serve as a proxy for the number of black engineering graduates from those schools in earlier years. If correct, the estimate that these account for at least half of all black engineering graduates will allow us to create an upper bound for all other institutions.

Information from different sources is consistent with the "at least half" estimate. The six HBCU6 institutions accounted for 60 percent of all black U.S. engineering graduates in each of the first two years of the EMC survey (Alden 1970, 1971). However, after correcting for a pair of typographical

15. This requirement was implemented shortly after the December 1980 passage of the federal Science and Technology Equal Opportunities Act (part B of Public Law 96-516).

16. Some earlier statistics exist. Wharton (1992) gathered names of black engineers who graduated between 1914 and 1929. He found about 400 (averaging twenty-seven per year), with the largest concentration of these from Howard University ($n = 36$, or 2.4 per year) and MIT ($n = 31$, or 2.1 per year). Early government statistics indicate that Howard University graduated thirteen engineers between 1922 and 1926 (2.6 per year), consistent with the counts made by Wharton (1992). Downing (1935) enumerated about 100 black engineering students in 1930, thirty-one at Howard University (where he was soon to become dean of the School of Engineering and Architecture), and the remainder in northern institutions including MIT, Cornell, Rensselaer Polytechnic Institute, Ohio State University, and the Universities of Michigan, Pennsylvania, Pittsburgh, Illinois, Wisconsin, and Armour Institute of Technology. Carey (1977) reports a count of 150 graduating black engineers in 1955.

errors that I discovered, the true statistic is closer to 65 percent.[17] The Project Talent longitudinal survey of students from the high school classes of 1960 and 1961 includes fourteen black students who held bachelor's degrees in engineering when they were resurveyed either five or eleven years after high school. Of the fourteen, all of those who graduated from southern high schools attended one of the HBCU6 institutions, and all students from northern high schools attended northern colleges. Reweighting the Project Talent data to control for the oversampling of certain high schools and students yields the estimate that 65 percent of the black engineers from these cohorts were educated at one of the HBCU6 campuses.[18] Another source of information is the 1993 National Survey of College Graduates (NSCG93), a retrospective survey that asked a representative sample of college graduates where and when they attended high school and college. While the data have been recoded to regional-level indicators to protect confidentiality, it is possible to estimate the proportion of black engineers educated during the 1960s who got degrees from colleges and universities in southern states. Based on a sample of thirty-seven respondents, I estimate that 63 percent of black engineers were educated in the South—most likely in HBCU6 programs. All of these suggest that "at least half" is an underestimate of the contribution of HBCU programs in the 1960s, and that the truth might be closer to two-thirds. Although imprecise, this estimate allows us to construct a proxy for national counts in earlier years, and helps us understand the degree to which growing EMC counts reflect true gains rather than more accurate counts.

The next step in the analysis is to compare growth in the number of black engineering graduates across different types of institutions. Figure 3.2 describes growth within different subsets of U.S. engineering programs using estimates in the earlier years from numbers recorded by the EMC, and later estimates from the U.S. Department of Education National Center for Education Statistics (NCES).[19] Figure 3.2, panel A, depicts trends in the number of black engineering graduates from HBCU campuses, with counts

17. The EMC warns that early year data may exaggerate racial stratification because many non-HBCU6 institutions left the racial counts blank. In fact, in the first year of the survey, more than half of the institutions left this question blank. Unless these blanks were all true zeros, they would lead to an exaggerated picture of the contribution of the HBCU6 institutions. As administrators became accustomed—eventually required—to answer these questions, the counts arguably got better. However, estimates based on other data sources suggest that most of the blanks are true zeros.

18. Although the sample is small, a 70 percent confidence interval has lower bound at 0.51, meaning there is only 15 percent probability that the true value is below half.

19. In the 1960s, I use the total number of engineering graduates from HBCU6 campuses to estimate black graduates from these campuses, and double this number to estimate the number of black graduates from all U.S. campuses. It is visually apparent that the estimates match up well in figure 3.2, panel A, but are mismatched in figure 3.2, panel B. Either HBCU6 campuses actually produced more than half in the 1960s, or the earliest EMC counts are too low due to nonreporting campuses.

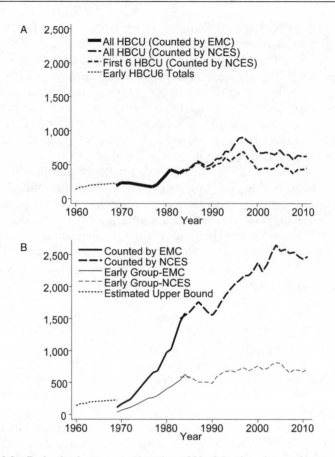

Fig. 3.2   By institution category, number of black engineering graduates from HBCU and other campuses, 1960–2011. *A*, HBCU campuses; *B*, Non-HBCU campuses.

Data source: Administrative records reported by educational institutions to the U.S. Department of Education or the Engineering Manpower Commission.

Note: Additional curve in panel A distinguishes between the original six HBCU engineering programs and new HBCU programs established later. Additional curve in panel B describes trends within a small subset of engineering programs that were early participants in efforts to increase the enrollment of black students.

for the original six institutions indicated separately from the full counts. It shows a sharp jump around 1980, reflecting the fact that between the mid-1970s and mid-1980s, the HBCU6 campuses expanded from an average of about 200 black graduates per year to about 500 graduates per year over a ten-year interval. Additional HBCU growth during the 1990s is primarily due to the opening of new HBCU engineering programs. Figure 3.2, panel B, depicts the corresponding counts for other U.S. campuses, and shows a steep upward trend in the number of black engineering graduates, eventually dwarfing the HBCU counts of figure 3.2, panel A.[20] The trend line labeled "Early Group" indicates subtotals for the thirty-four non-HBCU institutions with publicly stated early intention to change as recorded in the 1974 Blueprint Report (PCEMOE 1974).[21] This vanguard group—representing only 20 percent of U.S. engineering graduates—was responsible for a disproportionate 37 percent of the growth in black engineering graduates between 1969 and 1985.[22] Afterward, however, most of the growth was driven by expansion of black enrollment at additional institutions, as evidenced by the flattening of the "Early Group" line and continued growth of the total. A closer look at the EMC data suggests that the remaining institutions also made some initial changes by the mid-1980s: seventy percent of the 240 non-HBCU engineering schools outside the Early Group recorded at least one black graduate over the two-year interval covering 1983–1984 or 1984–1985, compared to only 30 percent over the first two years of EMC data collection.[23] These patterns suggest that early efforts led to small changes at a wide range of institutions and substantial growth in the number of black engineering graduates at a small number of institutions within a relatively short time frame, and spurred a movement that continued to expand in ensuing decades.

The timing and net effects of expansion were different in different geographic areas. Figure 3.3 combines the counts from all institutions and includes separate trend lines for large subsets of institutions, including those within the group of states containing the HBCU6 institutions, and the set of all institutions within southern states.[24] An additional trend line describes the unique contribution of Georgia Tech, which was averaging more than fifty black engineering graduates per year by the mid-1980s. The small difference between the trend lines for HBCU6 states and all southern states before 1980 indicates the initial dearth of opportunities for black students to major in engineering at southern institutions outside the HBCU6 locations.

The timing of changes in figure 3.3 suggests three distinct periods of expansion. During the early 1970s, expansion of opportunities to study engi-

20. These counts include graduates who began their studies in a dual-degree program on an HBCU campus. Additional research is required to estimate how many of these graduates participated in dual-degree programs. Close examination of NLSY cohorts that reached age twenty-two in 1979–1986 suggests that the number was small. Among black college graduates who spent any time as an engineering major, fewer than 10 percent could have earned dual degrees from an HBCU and another institution.

21. I included all non-HBCU institutions mentioned in the report in this category.

22. If Georgia Tech is excluded, the remaining members of the Early Group accounted for 33 percent of the total growth.

23. All of the engineering schools in the Early Group recorded at least one black graduate in either 1983–1984 or 1984–1985, compared to only 50 percent over the first two years of EMC data collection. Ninety percent of these, and 50 percent of programs outside the Early Group, averaged at least four per year by the mid-1980s.

24. The HBCU6 areas are Louisiana, North Carolina, Tennessee, Texas, Alabama, and the District of Columbia.

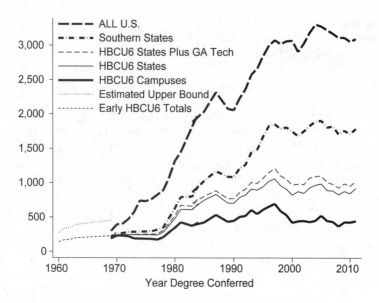

Fig. 3.3   Geographic breakdown of the number of black engineering graduates from U.S. campuses and selected subgroups, 1960–2011

Data source: Administrative records reported by educational institutions to the U.S. Department of Education or the Engineering Manpower Commission.

Note: Separate curves indicate counts for HBCU6 campuses (North Carolina A&T, Southern, Prairie View, Tennessee State, Tuskegee, and Howard), for HBCU6 states (North Carolina, Louisiana, Texas, Tennessee, Alabama, and the District of Columbia), for HBCU6 states plus Georgia Tech, and for all southern states.

neering was largely confined to (a handful of) northern institutions. In the late 1970s and early 1980s, a dramatic expansion occurred at the HBCU6 institutions, with some additional expansion at other institutions in the same six states. After this, the pattern of expansion spread to other southern states and to additional institutions in the north. These statistics indicate substantial change in a short period of time at both HBCU and other campuses, with particularly rapid per-institution impact at the HBCU6 institutions.

To better interpret the trends in numeric counts, figure 3.4 adjusts for simultaneous trends in the number of college graduates, describing how many black graduates (within various subsets of institutions) had engineering majors compared to the share among all U.S. students who graduated college the same year.[25] The topmost line indicates that before the intervention, new graduates of HBCU6 campuses were a bit more likely than the typical new U.S. graduate to hold an engineering degree, with a sharp

25. In figure 3.4, the state-level estimates from 1969–1972 rely on ballpark estimates of the total number of black college graduates in each region generated from information in the NSCG 1993 plus birth-cohort-specific information from the CPS. The short gap between 1973 and 1975 reflects years in which race-specific, institution-level data were not made public.

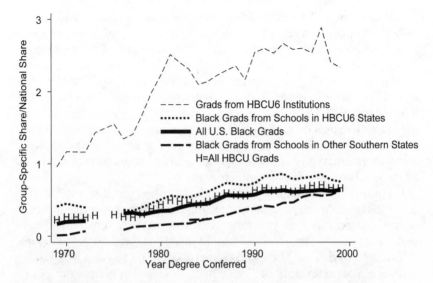

Fig. 3.4   Relative share of new graduates from each indicated group with an engineering major, based on counts of bachelor's degrees conferred in each year

Data source: Administrative records reported by educational institutions to the U.S. Department of Education National Center for Education Statistics or the Engineering Manpower Commission.

Notes: Relative shares are computed for each year as the proportion of graduates within indicated groups who majored in an engineering field divided by the proportion of all U.S. college graduates with an engineering major in the same year. The "HBCU6 States" category includes North Carolina, Louisiana, Texas, Tennessee, Alabama, and the District of Columbia, the locations of the original six black engineering campuses. Detailed counts were not made public between 1973 and 1975.

increase to more than twice as likely after the intervention. Below this, the bold line shows that the national relative share of black college students majoring in engineering followed a fairly smooth upward climb over three decades. The trend line indicated by the symbol "H" shows that as a group, the complete set of historically black colleges followed the U.S. trend most of the time, with the exception of a short-lived relative boost in the early 1980s. This pattern suggests that the main advantage of the HBCU campuses was in the ability of the HBCU6 to adjust quickly, rather than the total eventual amount of adjustment at HBCU relative to the much larger set of non-HBCU engineering campuses. The cost of acquiring engineering accreditation was too high for most HBCUs.

A geographic differential in the rate of expansion is also apparent in figure 3.4. The shorter dashed line shows that the upward trend came relatively sooner within the six southern states containing one of the HBCU6 campuses. The lowest curve underscores the lagged but persistent progress, eventually catching up to the national average, among southern campuses outside the HBCU6 states.

Although, overall, black students born in southern states with no historically black engineering campus lagged in the availability of new opportunities, there are notable exceptions. Since 1974, more than 4,000 black engineering students have graduated from Georgia Tech. More than one-third
of those graduates began their studies at an HBCU campus (Chubin, May,
and Babco 2005); in the earliest years more than one-half began at one of
four Atlanta HBCUs.[26] Thus, even in this case, HBCU campuses played an
important role as a conduit to facilitate the expansion of opportunities to
study engineering. Institutional and historical features of the educational
environment interacted with policy, leading to an uneven pace of expanded
educational access across geographic regions.

If students first choose a career path, and then travel to the location
where that path can be pursued, the geographic idiosyncrasies of the expansion of opportunities might not matter. But previous research suggests
that geographic distance from home and social distance matter to students
(Card 1995; Mykerezi, Mills, and Gomes 2003; Mykerezi and Mills 2008).
Research using data from the 1980s finds that geographic proximity to a
HBCU is a far better predictor of educational attainment among black
adults in a community than simple proximity to a college or university
(Mykerezi, Mills, and Gomes 2003). For this reason, differences in the timing of change across geographic regions, and between HBCU and other
campuses, are likely to have influenced which groups of young people were
affected by these changes in different years.

The Project Talent data described above suggested a strong geographic
component determining who was likely to attend the southern engineering
schools. This geographic pattern can also be seen in the far larger NSCG
sample. A sample of 331 African American engineering graduates drawn
from the NSCG93 data reveal that the vast majority of those who attended
high school in the South remained in the South during college, while most
of those from outside the South did not. When all cohorts of engineering
graduates are combined, 89 percent of those who graduated high school in
southern states attended college in the South ($n = 167$), and 88 percent of
those who graduated high school in northern states attended college in the
North ($n = 164$). Despite the proliferation of engineering recruitment efforts,
these patterns did not vary much over time. Among the younger cohorts
of this sample who graduated between 1976 and 1988, the corresponding
statistics are 90 percent ($n = 101$) and 87 percent ($n = 122$).[27] These statistics
further strengthen the case that the impacts of policy changes at HBCU6
campuses are likely to be geographically localized.

26. Estimated from data generously provided by Dr. Jane Weyant of Georgia Tech, combined
with EMC data. The Atlanta University Center-Georgia Tech dual degree program was established in 1969 with a grant from the Olin Charitable Trust Fund (Blackwell 1987).

27. Among those who earned engineering degrees between 1944 and 1965, 22 percent of
southern high school graduates ($n = 23$) and 96 percent of northern high school graduates
($n = 18$) attended college in the north.

Analysis of data linking place of birth to educational attainment confirms that those born near one of the HBCU6 institutions experienced particularly rapid increase in the propensity to study engineering. Using 2009–2013 ACS data for the full U.S.-born population, figure 3.5 shows that black students born in different parts of the United States were more likely to complete a bachelor's degree in engineering if they were born later, but that the rate of change within each region was related to the historical legacy of segregation and the geographic distribution of the HBCU6 institutions. Meanwhile, contemporaneous cohorts of black students from southern states with no historically black engineering campus enjoyed some gradual improvement, but experienced a persistent lag in the availability of expanded opportunities, relative to those in the HBCU6 states. On average, black students born outside the South had superior access to educational opportunities in engineering until the 1990s, when students from the HBCU6 states caught up. While figure 3.4 shows regional differences in the provision of education, figure 3.5 shows the consequences for black students born in different places and times.

Geographic differences in rates of change are robust to controls for local-

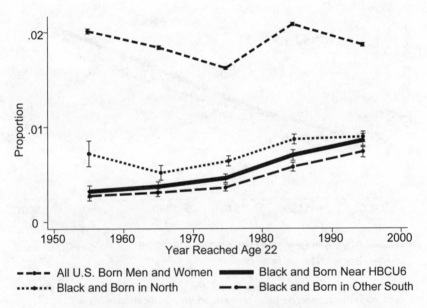

Fig. 3.5   Geographic differences in the propensity of black cohorts to graduate college with an engineering major by birth cohort and location of birth

Sample: American Community Survey 2009–2013, restricted to those born in the United States between 1928 and 1977 (age twenty-two in 1950–1999, divided into five ten-year birth cohorts), lower three curves restricted to those indicating black or African American heritage and birth in the indicated region of the country. The "Near HBCU6" category includes those born in North Carolina, Louisiana, Texas, Tennessee, Alabama, and the District of Columbia.

Note: Error bars represent 90 percent confidence intervals.

ized differences in rates of bachelor's degree attainment and the propensity of local college students to study engineering. Figure 3.6 presents ratios of ratios for each of the three regions: the proportion of black college graduates with an engineering degree divided by the proportion of all college graduates with an engineering degree in the same region and decade. The relative, conditional proportions displayed here accentuate the early and consistent success of the HBCU6 states in expanding opportunities for black college students to become engineers, and also highlights the lagged response of the southern states that initially lacked a historically black engineering college. As a group, these other southern states eventually caught up in the 1990s, possibly influenced by the benchmark set by the HBCU6 states. The more modest sustained relative gains in engineering-degree attainment among black college students from nonsouthern states are also evident here. While the national push to expand opportunities for black college students to become engineers had nationwide impacts, the timing and amount of expansion varied across regions of the country.

At the end of the period covered, parity had not been attained in engineer-

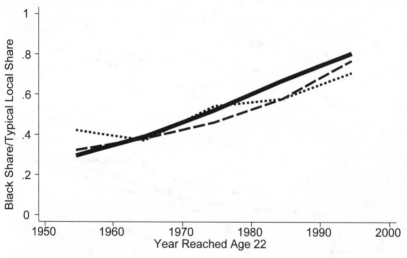

Fig. 3.6   Geographic differences in the relative propensity of black college graduates to major in engineering fields by birth cohort and location of birth

Sample: American Community Survey 2009–2013, restricted to college graduates born in the United States between 1928 and 1977 (age twenty-two in 1950–1999, divided into five ten-year birth cohorts).

Notes: Relative odds computed as the proportion of black college graduates with a degree in an engineering field divided by the proportion of all college graduates in the same region in the same decade with an engineering degree. The "Near HBCU6" category includes those born in North Carolina, Louisiana, Texas, Tennessee, Alabama, and the District of Columbia.

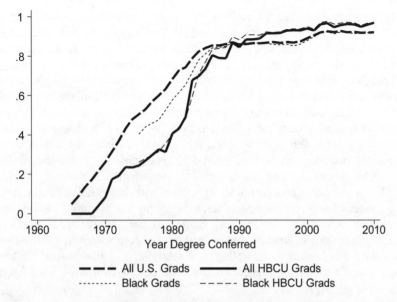

Fig. 3.7   Proportion of new graduates with computer or information science majors available at their institution for indicated subsets of graduates

Data source: Administrative records reported by educational institutions to the U.S. Department of Education.

ing in any of the three regions, even among college graduates. The share of black graduates with engineering degrees increased from about 40 percent of the national average to 70 or 80 percent, with the fastest increase attained near the HBCU6 locations.

3.2.2   Access to Computer Science Education

While nationwide efforts to broaden participation in engineering involved both HBCU and other campuses, a corresponding effort in computer sciences focused on HBCU campuses.[28] Between the 1970s and the early 1990s, as the number of historically black campuses with accredited engineering programs expanded from the original six to eleven, the number of HBCU campuses reporting new computer science graduates increased from a handful to more than sixty.

Computer science did not emerge as a popular and available college major choice until the late 1970s. Figure 3.7 describes the spread of opportunities to study computer science across the country as more U.S. campuses began to offer this major. In 1965, very few students of any race had opportunities to major in computer science. By the mid-1970s only half of all U.S. college

28. A conversation with John Brooks Slaughter (2012), who headed NSF at the time, confirmed that this change was driven by the administrations of the colleges and universities themselves, and was not a "top-down" effort.

students, and 40 percent of black college students, graduated from a campus where a computer science major was offered; among those black students at HBCU campuses, only a quarter had access to a computer science course of study at that time. These racial gaps persisted until the mid-1980s, when very rapid changes at HBCU campuses completely eliminated this differential. In 1980, new HBCU college graduates were only 60 percent as likely as new U.S. graduates to have classmates with a computer science major, by 1990 new HBCU graduates were just as likely as the typical new U.S. graduate to have classmates who were computer science majors. The introduction of computer science programs to additional HBCU campuses entirely accounts for the disappearance of the gap in access to computer science courses.

This increase in access quickly translated into increased attainment of computer science degrees among black college graduates (figure 3.8). In contrast to the situation in engineering (figure 3.4), policies at HBCU campuses during the mid-1980s promulgated the choice of computer and information science majors among black college students relative to all U.S. students and also relative to black college students at other U.S. campuses. This figure

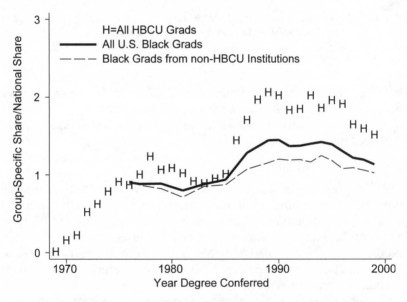

Fig. 3.8   Relative share of new graduates with computer or information science majors based on counts of bachelor's degrees conferred in each year for indicated subsets of institutions or students

Data source: Administrative records reported by educational institutions to the U.S. Department of Education.

Note: Relative shares are computed for each year as the proportion of graduates within indicated groups who majored in a computer or information science field divided by the proportion of all U.S. college graduates with a computer or information science major in the same year.

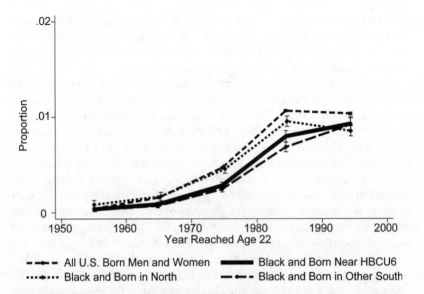

Fig. 3.9   Geographic differences in the propensity of black cohorts to graduate college with a computer science major by birth cohort and location of birth

Sample: American Community Survey 2009–2013, restricted to those born in the United States between 1928 and 1977 (age twenty-two in 1950–1999, divided into five ten-year birth cohorts), lower three curves restricted to those indicating black or African American heritage and birth in the indicated region of the country. The "Near HBCU6" category includes those born in North Carolina, Louisiana, Texas, Tennessee, Alabama, and the District of Columbia.

Note: Error bars represent 90 percent confidence intervals.

shows very rapid growth during the 1970s in the share of HBCU graduates with computer science majors. While there are no corresponding data on black computer science graduates at other campuses until the late 1970s, the information that does exist suggests that early HBCU policies played an important role in preventing a large racial gap in the propensity to study computer science from ever emerging. By the mid-1970s, new HBCU graduates were as likely as all U.S. graduates to hold a computer science degree. By the late 1980s, the same was true of new black graduates from other U.S. institutions, and HBCU graduates were majoring in computer science at a rate twice as high as the national average.

The result is that, in contrast to the large racial differential in engineering (see figure 3.5), the differential in computer and information sciences is quite small as observed in the ACS 2009–2013 data (see figure 3.9). Black cohorts of all ages are nearly as likely to hold computer science degrees as the U.S. average, in contrast to the large and persistent racial gaps in engineering attainment. Even as a share of the full population reaching age twenty-two in the 1990s regardless of education, African Americans were 74 percent as likely to attain a computer science degree, and 86 percent as likely to attain a degree in a more broadly defined set of computer and information systems majors, compared to being 45 percent as likely to attain a degree in engineering. Figure 3.9 also shows that, compared to engineering, the remaining gap is small in both southern and other states.

As figure 3.8 already made clear, the remaining racial gap in computer science degree attainment is entirely due to differences in the propensity to attain a college degree. Among all college graduates who reached age twenty-two in the 1970s, 1980s, or 1990s, African Americans were *more likely* than others to major in computer science fields. In sum, the localized effort centered at HBCU campuses to expand computer sciences led to strong participation of black college students in computer science from the outset of the field far in excess of their relative participation in engineering. But, did the increase in computer science and engineering majors lead to a net increase in the total number of STEM majors, or did it represent the transfer of students from one STEM field to another? Figure 3.10 shows the relative shares of new black graduates with majors in engineering, computer science, and also in either of these two majors *or* math or other sciences. The slopes indicate similar rates of change for engineering, com-

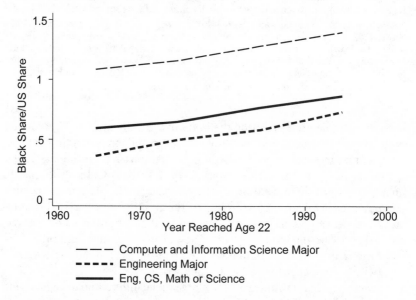

Fig. 3.10   Nationwide trends in the relative odds that a black graduate has a computer science, engineering, math, or other science major compared to the typical likelihood among U.S. college graduates

Sample: American Community Survey 2009–2013, restricted to those born in the United States between 1938 and 1977 (age twenty-two in 1960–1999, divided into four ten-year birth cohorts).

Note: Relative odds computed as the share of black cohort members with indicated major divided by the share of all U.S. cohort members with the same major.

puter science, and the all-STEM category. There was no trend in the "math and other sciences" category alone, which was 74 percent in both the 1960s and 1990s (not depicted).[29] Overall, relative STEM participation grew from 59 percent to 86 percent over this period. The overrepresentation of black students in computer science majors, particularly on HBCU campuses, can be credited with pulling the STEM average closer to parity among college graduates, despite persistent underrepresentation in engineering and other STEM fields.

3.2.3   The Entry of Black Women to Engineering and Computer Science

Gender was an afterthought in national efforts to expand minority involvement in engineering careers. The "Blueprint for Action" barely mentions women at all. Initially, the expansion of opportunities just described largely affected men.[30] However, there was a rapid expansion of opportunities for black women to study engineering during the late 1970s and early 1980s. During the early 1960s the six original HBCU engineering campuses graduated a total of about two women per year between them, rising to four or five by the end of the 1960s. By 1980, more than one hundred women per year were completing degrees at the HBCU6 campuses, and most of these women were black. Within other campuses, there is no way to estimate the number of black women engineering graduates at the institution level in earlier years, though the number was almost certainly small.[31] Due to the absence of data for earlier years, only the share of all engineering graduates who were women (regardless of race), or the share of all engineering graduates who were black (regardless of gender), can be computed consistently at the institution level over a longer interval.

To the extent possible, given data limitations, it is important to understand the extent to which access to engineering education expanded for black women as well as men. Figure 3.11 displays the time trends in the proportion of new engineering graduates who were women among all engineering graduates in the United States, and among HBCU graduates. In each year, the historically black engineering colleges graduated a higher share of women than the typical U.S. engineering college. In 1970 the difference was relatively small, and very few women studied engineering at that time. During the 1970s and 1980s, the share of all U.S. engineering graduates who were

---

29. There was a transitory dip to 65 percent in between, due to increase in the reference group rather than decrease among black graduates.

30. For a historical perspective on women's entry to engineering, see Bix (1999, 2004, 2013). Biographies of Julia Morgan, whose architecture survived severe California earthquakes due to her early training as a civil engineer, offer some insight into the educational environment during an earlier era.

31. Separate counts for black women engineering graduates were not collected until the 1981–1982 academic year, although some earlier years counted the combined total of black women in either engineering or engineering technology fields, placing an upper bound on the number in engineering.

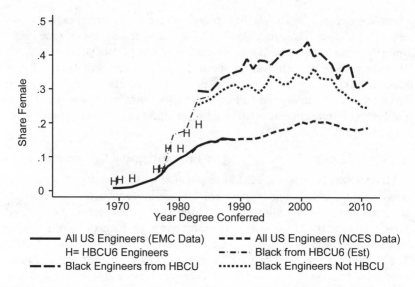

Fig. 3.11 Share female among new engineering graduates, 1969–2011, all U.S. and selected subgroups

Data source: Administrative records reported by educational institutions to the U.S. Department of Education or the Engineering Manpower Commission.

women increased rapidly, flattening out in the late 1980s. However, at the historically black engineering colleges, the share of engineering graduates who were women grew much more rapidly than the national average, and did not flatten out until surpassing the national average by a factor of two. For the later years, estimates of the relatively high share of black engineering graduates who are female are also displayed; by the first time detailed statistics were collected, more than a quarter of black engineering students were women. These indicate high national ratios of women to men among black engineering graduates on average as well as at HBCU campuses.

As the EMC data covering the 1970s do not break down statistics by both race and gender, we cannot determine the full picture for these years. However, an estimate based on all black students in either engineering or engineering technology fields at the HBCU6 campuses shows rapid change. Patchy evidence from the NSCG reveals that, among black engineering graduates who completed bachelor's degrees between 1960 and 1974, only 6 percent of new engineering graduates who attended college in the South ($n = 39$), and none of those who attended college outside the South ($n = 39$), were women. By the 1975 to 1979 graduation window, the share female among black engineering graduates had risen to 13 percent ($n = 26$) in the South, and 3 percent outside the South ($n = 27$), and by 1980 to 1984 to 27 percent ($n = 59$) in the South and 33 percent outside the South ($n = 57$). Going back further, early U.S. Office of Education statistics on the number of women graduating from HBCU6 engineering programs show zero recorded between 1952 and 1956, and an average of two per year (0.33 per campus per year) between 1957 and 1959. Taken together, these statistics indicate that the high ratio of women to men among black engineering graduates observed in the 1980s and later is not the continuation of an older pattern, but likely the result of new policies that dramatically expanded opportunities for black women to study engineering at both HBCU and other campuses.

As in engineering, the representation of women among computer and information science graduates is higher among black graduates, and among HBCU graduates, than in the full population. Unlike engineering, data broken down by race and gender are available beginning with the 1975–1976 academic year, and the strong relative participation of black women, especially at HBCU campuses, seems to have begun many years before large numbers of black women were pulled into engineering. Over the most recent decade, the share of computer and information science degrees going to women of all races has trended downward. Nonetheless, in each of the past four decades black women have been a strong presence among computer and information science graduates. The contrast between figure 3.11, showing the transition toward increasing entry of black women to engineering, and figure 3.12, showing the strong participation of women educated at historically black colleges from near the inception of the computer science

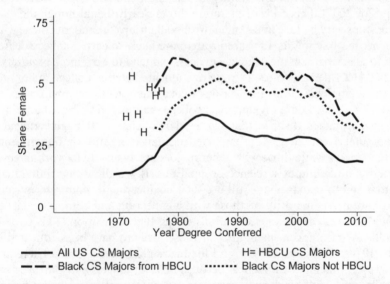

Fig. 3.12   Share female among new computer and information science graduates, U.S. and selected subgroups

Data source: Administrative records reported by educational institutions to the U.S. Department of Education.

field, is striking. In both cases, inclusion of women in engineering and computer science majors is clearly an important factor in the changes that occurred at HBCU campuses during this period.

3.3 Labor Market Outcomes

Recall that the initial motivation for policies to bring black students into engineering and computer science majors was to prepare them for careers not only in these technical fields, but also as future industrial leaders. As Royster's (2003) study of early career experiences among equally well-trained graduates of a public vocational high school shows, the link from expanded educational opportunity to career cannot be taken for granted, even in the post-civil-rights era. The evidence presented in the following section shows that, on average, black workers who graduated college with majors in engineering and computer sciences during the 1970s, 1980s, or 1990s are working in well-paid technical or managerial occupations in today's labor market.

A long literature documents correlations between high school math scores, entry into engineering or computer science college majors, and adult earnings (Fiorito and Dauffenbach 1982; Blakemore and Low 1984; Paglin and Rufolo 1990; Benbow and Arjmand 1990; Murnane, Willet, and Levy 1995; Grogger and Eide 1995; Brown and Corcoran 1997; Weinberger 1998, 1999, 2001; Turner and Bowen 1999; Xie and Shauman 2003; Weinberger and Joy 2007). This body of literature shows clearly that although the relationships vary by race and gender, within all groups of students those with higher math scores as adolescents are more likely to earn a college degree or to select mathematical college majors, and tend to earn more as adults.

Table 3.1 demonstrates a strong relationship between college major and current employment in engineering or computer science occupations using 2009–2013 ACS data on employed college graduates holding a bachelor's degree or higher. It shows that a large share of those with engineering and computer science majors are employed in engineering and computer science occupations, while those with other majors are less likely to work in engineering and computer science occupations. If we allow that individuals promoted to management still use their engineering or computer science education, the majority of those with engineering and computer science majors appear to be in occupations fitting their education. This pattern holds for men and women, and for those likely to have been educated in the 1970s, 1980s, and 1990s, and for black graduates as well. There are, however, systematic differences between black graduates and U.S. averages in the table, with black engineering or computer science graduates having moderately smaller likelihoods of being employed in engineering, computer science, or management careers. Because these jobs are so highly correlated with college major, there are surely more black engineers and computer

Table 3.1 Percentage of indicated groups of employed college graduates reporting engineering or computer and information science occupation in 2009–2013

	(1) Men age 22 in 1970–1979, observed 2009–2013	(2) Men age 22 in 1980–1989, observed 2009–2013	(3) Men age 22 in 1990–1999, observed 2009–2013	(4) Women age 22 in 1970–1979, observed 2009–2013	(5) Women age 22 in 1980–1989, observed 2009–2013	(6) Women age 22 in 1990–1999, observed 2009–2013
Engineering majors in eng. or CS occupations	44	47	49	32	37	39
Computer majors in eng. or CS occupations	53	55	60	38	39	39
Math & statistics majors in eng. or CS occupations	26	31	27	16	19	15
Biological & physical sciences majors in eng. or CS occupations	9	10	9	4	4	3
Other majors in eng. or CS occupations	5	6	7	2	2	2
Engineering majors in eng. or CS or managerial occupations	67	71	70	51	59	59
Computer majors in eng. or CS or managerial occupations	67	70	72	51	54	52
Black engineering majors in eng. or CS occupations	44	40	43	32	29	32
Black computer majors in eng. or CS occupations	40	45	51	32	30	39
Black math & statistics majors in eng. or CS occupations	21	27	16	13	20	14
Black biological & physical sciences majors in eng. or CS occupations	6	8	7	3	3	3
Other black majors in eng. or CS occupations	4	5	5	2	2	2
Black engineering majors in eng. or CS or managerial occupations	61	60	62	50	55	53
Black computer majors in eng. or CS or managerial occupations	51	57	60	43	45	50

professionals today than there would have been in the absence of policies to expand educational opportunities in these fields.

What about earnings premiums for black graduates in these fields? It seems likely that the same policies that increased the number of black students entering these majors might have led to changes in the associated earnings, growing over time if factors such as the quality of instruction, precollege preparation, or professional networks of successive graduating classes improved over time, or declining over time if the number of new black engineering graduates grew more quickly than the number of less discriminatory employers. While other between-cohort changes besides those pertaining to engineering or computer science education almost certainly affected the relative earnings of black workers in different age groups, I focus solely on the relation of earnings to college majors.[32]

Table 3.2 presents earnings regressions for the entire current U.S. workforce that estimate log earnings differentials for engineering, computer science, and other science or mathematics majors compared to the omitted group of high school graduates. These regressions do not include measures of occupation, so the estimated earnings boost associated with college major includes the effects of major on the set of attainable occupations, as depicted in table 3.1, and differences in earnings within occupations as well as between them. Column (1) combines men and women of all three birth cohorts, columns (2)–(4) describe the oldest, middle, and youngest cohorts of men, and columns (5)–(7) describe the three cohorts of women. The regressions show consistently high earnings for engineering, computer science, and other science majors, both relative to high school graduates born in the same state and year, and relative to other college graduates. Across all three age groups, both men and women with nonscience bachelor's degrees earn 0.5–0.6 log points more, on average, than the typical high school graduate the same age and from the same birth state. Among men who majored in engineering, there is an additional 0.2–0.3 log points earnings premium, for a total premium near 0.8 log points relative to high school graduates (columns [2]–[4]).

---

32. Chay, Guryan, and Mazumder (2009, 2014) present evidence that improved access to healthcare during infancy and early childhood was associated with improved health and rising test scores—especially at the upper tail of the math score distribution—among black students in the south. This shift primarily affected the latest cohort in our analysis, those who turned twenty-two during the 1990s or later. Other policy changes during this era, including school desegregation orders, improved K–12 education in southern states, and government programs to improve health, nutrition, and access to preschool education have been linked to positive impacts on health, educational attainment, and earnings among later cohorts (Card and Krueger 1992, 1996; Chay and Greenstone 2000; Garces, Thomas, and Currie 2002; Currie and Moretti 2008; Johnson 2010, 2011; Almond, Hoynes, and Schanzenbach 2011; Hoynes, Schanzenbach, and Almond 2016). The later cohorts also enjoyed increased access to courses in Algebra II during high school (Weinberger 2014). Despite these positive changes for cohorts who would reach adulthood in the 1980s and later, labor market opportunities for young black workers completing school improved only from the mid-1960s through the late 1970s, and then declined, particularly among college graduates (Bound and Freeman 1992).

Table 3.2 Log annual earnings premiums to indicated educational attainment categories, all U.S. full-time, full-year workers, by birth cohort and sex

	(1) All U.S. age 22 in 1970–1999, observed 2009–2013	(2) Men age 22 in 1970–1979, observed 2009–2013	(3) Men age 22 in 1980–1989, observed 2009–2013	(4) Men age 22 in 1990–1999, observed 2009–2013	(5) Women age 22 in 1970–1979, observed 2009–2013	(6) Women age 22 in 1980–1989, observed 2009–2013	(7) Women age 22 in 1990–1999, observed 2009–2013
Categories of college graduate:							
Engineering major	0.821	0.785	0.818	0.775	0.898	1.001	0.993
	(0.003)***	(0.006)***	(0.005)***	(0.005)***	(0.026)***	(0.014)***	(0.013)***
Computer major	0.714	0.613	0.710	0.683	0.735	0.823	0.779
	(0.004)***	(0.011)***	(0.007)***	(0.007)***	(0.024)***	(0.011)***	(0.015)***
Other science or math	0.741	0.724	0.746	0.686	0.736	0.793	0.777
	(0.003)***	(0.007)***	(0.007)***	(0.007)***	(0.009)***	(0.008)***	(0.008)***
Other major	0.550	0.500	0.566	0.543	0.521	0.580	0.579
	(0.001)***	(0.004)***	(0.003)***	(0.003)***	(0.003)***	(0.003)***	(0.003)***
Higher degree	0.235	0.260	0.243	0.234	0.244	0.220	0.214
	(0.002)***	(0.004)***	(0.004)***	(0.004)***	(0.004)***	(0.004)***	(0.004)***
Some college	0.207	0.188	0.206	0.200	0.229	0.226	0.202
	(0.001)***	(0.003)***	(0.002)***	(0.003)***	(0.003)***	(0.002)***	(0.003)***
Less than HS	−0.215	−0.191	−0.218	−0.197	−0.239	−0.250	−0.229
	(0.002)***	(0.005)***	(0.005)***	(0.006)***	(0.006)***	(0.006)***	(0.007)***
Female	−0.325						
	(0.001)***						
Black male	−0.223	−0.232	−0.247	−0.215			
	(0.002)***	(0.004)***	(0.004)***	(0.004)***			
Black female	−0.100				−0.081	−0.095	−0.091
	(0.002)***				(0.004)***	(0.003)***	(0.003)***
Observations	2,708,875	448,416	574,087	474,051	378,723	472,681	360,917
R^2	0.29	0.23	0.26	0.25	0.24	0.26	0.28

Notes: Dependent variable: natural log of annual income, inflation-adjusted to 2010 dollars; sample: U.S.-born full-time, full-year workers from the 2009–2013 American Community Survey, with annual income observed and greater than $2,000; additional controls: fixed effects for year-specific birth cohort × state of birth (or D.C.), for a total of 510 categories per regression.

All estimates reported are national averages relative to the typical high school graduate from the same birth cohort and same birth state, with the exception of "higher degree," which is an average additional premium relative to the typical college graduate with the same undergraduate college major.

***Significant at the 1 percent level.

**Significant at the 5 percent level.

*Significant at the 10 percent level.

Among women who majored in engineering, the additional premium above the earnings of nonscience college graduates is even higher, closer to 0.4 log points above the earnings of women the same age and same state of birth with nonscience college majors, for a total premium more than double the earnings of high school graduates (columns [5]–[7]). The earnings of computer science and other science majors (including physical sciences, biological sciences, mathematics, and statistics) are not quite as high as those of engineering majors, but tend to be substantially higher than the earnings of nonscience college majors.

Table 3.2 also includes estimates of gender and racial wage differentials. The regression of column (1) indicates that when all three cohorts are combined, women earn 0.3 log points less than men on average. Black women face an additional 0.1 log point disadvantage, for a total disadvantage of 0.4 log points. Black men face a somewhat smaller but still substantial 0.2 log point disadvantage. These differentials are large, and are similar across cohorts, fluctuating with no apparent trend. However, these average differentials cannot tell us how each new cohort of black engineers and computer scientists fared, relative to other black college graduates or black high school graduates born in the same state and year.

Tables 3.3 and 3.4 show regressions that are comparable to those in table 3.2, with the samples restricted to black men and women, and showing separate specifications for those born in southern versus other U.S. states. This division is motivated by the concentration of HBCU campuses in the southern states, and the faster pace of change in the number of black engineering and computer science graduates in southern states during this time. Table 3.3 gives estimated earnings differentials for men, while table 3.4 gives the differentials for women. In both tables column (1) relates to the oldest group in the ACS sample—those educated in the 1970s; column (2) relates to the group educated in the 1980s, while column (3) relates to the youngest group—those educated in the 1990s.

The estimates in column (1) of tables 3.3 and 3.4 are for black men and women likely to have been educated in the North during the 1970s. In these groups, the earnings premiums to engineering or computer science degrees are low compared to the U.S. averages shown in table 3.2. We lack data to assess whether these groups received lower-quality education or simply faced higher barriers to obtaining early career employment commensurate with their education. Insights from previous research include the observation that black graduates of HBCU campuses earned more than graduates of other institutions among those educated in the 1970s (Constantine 1995), but are not as well compensated among those educated more recently (Fryer and Greenstone 2010). One hypothesis is that other institutions became better at educating black students over time (Fryer and Greenstone 2010). Another possible explanation is that employers inclined to hire black graduates searched primarily at HBCU campuses during the 1970s but no

Table 3.3 Log annual earnings premiums to indicated educational attainment categories, black male full-time, full-year workers, by birth cohort and birth region

	Born in northern states			Born in southern states		
	(1) Black men age 22 in 1970–1979, observed 2009–2013	(2) Black men age 22 in 1980–1989, observed 2009–2013	(3) Black men age 22 in 1990–1999, observed 2009–2013	(4) Black men age 22 in 1970–1979, observed 2009–2013	(5) Black men age 22 in 1980–1989, observed 2009–2013	(6) Black men age 22 in 1990–1999, observed 2009–2013
Categories of college graduate:						
Engineering major	0.601	0.880	0.820	0.847	0.747	0.827
	(0.096)***	(0.040)***	(0.033)***	(0.048)***	(0.043)***	(0.038)***
Computer major	0.496	0.701	0.644	0.694	0.778	0.706
	(0.087)***	(0.048)***	(0.039)***	(0.053)***	(0.036)***	(0.034)***
Other science or math	0.672	0.727	0.801	0.698	0.787	0.661
	(0.057)***	(0.064)***	(0.057)***	(0.049)***	(0.050)***	(0.045)***
Other major	0.462	0.499	0.491	0.459	0.512	0.480
	(0.027)***	(0.018)***	(0.016)***	(0.017)***	(0.016)***	(0.014)***
Higher degree	0.327	0.243	0.277	0.247	0.223	0.232
	(0.032)***	(0.025)***	(0.023)***	(0.024)***	(0.023)***	(0.021)***
Some college	0.213	0.201	0.202	0.197	0.233	0.217
	(0.018)***	(0.013)***	(0.013)***	(0.011)***	(0.010)***	(0.010)***
Less than HS	−0.172	−0.242	−0.232	−0.167	−0.221	−0.197
	(0.032)***	(0.026)***	(0.026)***	(0.015)***	(0.016)***	(0.019)***
Observations	9,635	17,692	17,432	22,215	28,223	22,299
R^2	0.21	0.20	0.22	0.18	0.19	0.21

Notes: Dependent variable: natural log of annual income, inflation-adjusted to 2010 dollars; sample: U.S.-born full-time, full-year workers from the 2009–2013 American Community Survey, with annual income observed and greater than $2,000, restricted to men who identified themselves as black or African American; additional controls: fixed effects for year-specific birth cohort × state of birth (or D.C.). All estimates reported are averages relative to the typical black male high school graduate from the same birth cohort and same birth state, with the exception of "higher degree," which is an average additional premium relative to the typical black male college graduate with the same undergraduate college major.

***Significant at the 1 percent level.

**Significant at the 5 percent level.

*Significant at the 10 percent level.

Table 3.4 Log annual earnings premiums to indicated educational attainment categories, black female full-time, full-year workers, by birth cohort and birth region

	Born in northern states			Born in southern states		
	(1) Black women age 22 in 1970–1979, observed 2009–2013	(2) Black women age 22 in 1980–1989, observed 2009–2013	(3) Black women age 22 in 1990–1999, observed 2009–2013	(4) Black women age 22 in 1970–1979, observed 2009–2013	(5) Black women age 22 in 1980–1989, observed 2009–2013	(6) Black women age 22 in 1990–1999, observed 2009–2013
Categories of college graduate:						
Engineering major	0.570	0.932	0.860	1.195	1.041	0.955
	(0.188)***	(0.061)***	(0.044)***	(0.083)***	(0.069)***	(0.049)***
Computer major	0.433	0.706	0.635	0.640	0.766	0.812
	(0.191)**	(0.037)***	(0.046)***	(0.064)***	(0.038)***	(0.039)***
Other science or math	0.639	0.775	0.777	0.875	0.765	0.769
	(0.057)***	(0.049)***	(0.048)***	(0.050)***	(0.050)***	(0.034)***
Other major	0.513	0.531	0.469	0.545	0.589	0.546
	(0.020)***	(0.014)***	(0.014)***	(0.013)***	(0.011)***	(0.011)***
Higher degree	0.239	0.241	0.242	0.234	0.219	0.201
	(0.023)***	(0.017)***	(0.014)***	(0.016)***	(0.015)***	(0.013)***
Some college	0.182	0.204	0.162	0.238	0.253	0.195
	(0.015)***	(0.012)***	(0.013)***	(0.009)***	(0.008)***	(0.010)***
Less than HS	−0.217	−0.216	−0.188	−0.200	−0.235	−0.229
	(0.033)***	(0.025)***	(0.028)***	(0.015)***	(0.015)***	(0.020)***
Observations	12,420	21,539	21,405	28,477	35,894	28,379
R^2	0.24	0.23	0.24	0.25	0.26	0.28

Notes: Dependent variable: natural log of annual income, inflation-adjusted to 2010 dollars; sample: U.S.-born full-time, full-year workers from the 2009–2013 American Community Survey, with annual income observed and greater than $2,000, restricted to women who identified themselves as black or African American; additional controls: fixed effects for year-specific birth cohort × state of birth (or D.C.). All estimates reported are averages relative to the typical black female high school graduate from the same birth cohort and same birth state, with the exception of "higher degree," which is an average additional premium relative to the typical black female college graduate with the same undergraduate college major.

***Significant at the 1 percent level.

**Significant at the 5 percent level.

*Significant at the 10 percent level.

longer do so. Consistent with this possibility, Freeman (1976a) reported a dramatic increase in the number of recruiters sent to the thirty black campuses in his study, from an average of four per school in 1960 to more than 300 in 1970. Contemporaneous advice to corporate recruiters recommended that it is better to go directly to HBCU campuses, rather than risk stirring up tensions by recruiting minority students at a recently integrated campus (Lusterman 1979). If the recruiters with the most inclination to hire black graduates tended to avoid northern campuses, the group of graduates educated in the North during the 1970s might have had less favorable early work experience.

In contrast, the coefficients in columns (2)–(6) of table 3.3 show earnings premiums to engineering, computer information science, and other science majors that are comparable in magnitude to the corresponding premiums for all US persons in table 3.2. In most cases, relative to high school graduates, earnings premiums for engineering range from 0.75 to 1.0, compared to 0.78 to 1.0 for all U.S. men and women, and premiums to computer science and other sciences range from 0.6 to 0.9, compared to 0.6 to 0.8 for all U.S. men and women.[33] These estimates are quite close despite the somewhat lower match rate to engineering and computer science occupations. There is no evidence that the labor market value of an engineering or computer science degree declined as larger shares of newer cohorts of the black population entered these fields. It is noteworthy that the engineering and computer science coefficients in columns (4) and (5) of both tables 3.3 and 3.4 (southern cohorts entering the labor market in the 1970s vs. the 1980s) are relatively stable. These estimates span the largest increase in numbers of black college graduates with majors in engineering or computer science, and involve cohorts that predate any changes in the national distribution of math test scores among high school seniors.[34]

Overall, the data suggest that despite rapid changes in educational opportunities throughout the 1970s, 1980s, and 1990s, the current labor market places a high average value on black graduates who earned degrees in engineering or computer science fields during that time, relative to those who chose other educational paths.[35]

33. The estimated engineering premium to the oldest group of nineteen southern women is higher than the estimates for later cohorts, but is not statistically different from the estimate for the adjacent cohort.

34. See the NCES (1994) publication "NAEP Trends in Academic Progress" for evidence on the timing of these trends. Goodman (2017) documents increases in high school mathematics course taking that most likely affected cohorts graduating college in the 1990s and later.

35. Previous research found that, among younger college graduates, the earnings premium to engineering and computer science majors is larger for black than for white students, partially mitigating the overall racial gap in earnings (Freeman 1976a; Weinberger and Joy 2007). This does not appear to be the case in this older sample.

3.4   Conclusion

Nationwide, the racial differential in the propensity to become an engineer has narrowed continuously since the 1970s but with different rates and timing across states. Six historically black engineering programs played an important early role in facilitating change. The research in this chapter shows that when the nation decided to invest in expanded opportunities for black students to become engineers, the six HBCU engineering campuses responded more quickly than most other campuses. A few years later, a broader coalition of HBCU campuses dramatically expanded opportunities to study computer and information sciences near the inception of the field, effectively preventing a large gap in the national number of black computer science graduates from ever becoming the status quo. As a unique component of the U.S. educational infrastructure, HBCU campuses continue to play an important role in our economy, providing a conduit to broaden participation in engineering, computer science, and other technical careers. The graduates of these and other campuses fill valued roles in our current engineering and computer science workforce, and tend to earn substantially more than college graduates with other majors.

It is sometimes difficult to discern whether a targeted minority engineering program actually increases participation or simply reshuffles capable students from one program to another. My analysis suggests that expansion of engineering programs at six HBCU campuses led to localized impacts on the entry of black students to engineering careers. The timing of responses to this "natural experiment" indicates that the geographic distribution of educational opportunities can influence important outcomes. The larger lesson of this inquiry is that educational policies can influence the future career paths of students and the supply of our educated workforce.

Data Appendix

American Community Survey (ACS)

All regressions, figures 3.1, 3.5–3.7, 3.9, and 3.10, and statistics for birth cohorts described by the year of reaching age twenty-two are based on data from the 2009–2013 American Community Survey, collected by the U.S. Bureau of Labor Statistics and provided as part of the IPUMS Project (Ruggles et al. 2010). The sample is restricted to individuals at least thirty-one years old because this group is likely to have completed their education. Because the focus is on the U.S. educational system, individuals born in other countries are not included in the sample. Individual college graduates are coded as holding a degree with an engineering or computer science major based on either the first or second bachelor's degree major indicated. A narrow computer science category is limited to the exact "computer science" designation. A broad computer science category including all majors under the subheading "computer and information systems" plus the interdisciplinary major "math and computer science" was used throughout the chapter, and the engineering designation was similarly inclusive. "Black" is defined as indicating black or African American heritage. The "college graduate" category includes all bachelor's degree graduates, including those who also hold higher degrees.

In the earnings regressions, college major categories are mutually exclusive; the engineering category includes everyone with first or second major listed as engineering; computer science includes those with first or second major listed in a computer or information systems field, but no engineering major listed; and other science or math includes those with first or second major in physical sciences, biological sciences, mathematics or statistics, but no engineering or computer major listed. All remaining college graduates, and all graduates with college major "allocated" by the BLS, are coded to the "other major" category in the earnings regressions. In the figures, allocated majors are treated at face value.

All estimates are weighted by the person-specific weight.

U.S. Department of Education National Center for Education Statistics (NCES)

Institution-level counts of the number of bachelor's degrees conferred by each institution, by gender and academic field of study, were collected by the U.S. Department of Education under the HEGIS program in earlier years, and IPEDS beginning in 1982–1983. These data are currently maintained by the National Center for Education Statistics (NCES). These counts represent a (nearly) complete census of all bachelor's degrees conferred in the United States each year. Statistics broken down by race were collected in 1975–1976, then every other year between 1976–1977 and 1988–1989, and every year since that time. The disadvantage of this data source is that, in the earlier years, the counts by race combine engineers with bachelor's degree graduates who majored in engineering technologies. Therefore, an additional source is used to fill in more detailed information about black engineering graduates before 1983.

For the decade before any racial counts were systematically collected, the total number of engineering degrees conferred by HBCU campuses is used as an estimate of the number of black graduates from those campuses. This number also serves as an upper-bound estimate of the number of black engineering graduates from all remaining U.S. campuses, based on estimates presented in the chapter and elsewhere that HBCU campuses produced at least half of all black engineering graduates at the time.

Information about institutional characteristics was also collected. Where possible, these were aggregated to create a set of time-invariant indicators.

In the original surveys, the level of aggregation across components of a given institution was not uniform over time. To facilitate within-institution (or within-state) comparisons over time, time-consistent institution groupings were constructed. A particular institution grouping might include all campuses of a larger system (if that system ever reported statistics for all campuses combined), or might include two institutions that merged together at some point in time, or a pair of institutions that later split into smaller units for reporting statistics. Graphs depicting time trends within a group of institutions, or within a set of states, include a constant, completely comparable set of institution groupings in each year's estimate.

Care was also taken to maintain comparable definitions of "engineering majors" and "computer science majors" over time. Engineering does not include engineering technology degrees, but is otherwise inclusive of different fields of engineering. In cases where a subfield was categorized as engineering technology in some years and engineering in other years, the most time-consistent assignment was chosen.

Engineering Manpower Commission (EMC)

Beginning in the late 1960s, the Engineering Manpower Commission (EMC) collected similar institution-level counts of the number of bachelor's-level engineering graduates, with more detailed information about field of degrees in engineering or engineering technologies. These surveys include earlier counts by race than the NCES data, but the racial counts are not disaggregated by sex. The advantage of this data source in the early years is that black engineers are counted separately from black bachelor's degree graduates with majors in engineering technologies.

These data were not available in electronic form. They were collected from published volumes and matched to the NCES institution groupings by institution name (Tolliver and Armsby 1959; Tolliver, Armsby, and United States 1961, 1963; Engineering Manpower Commission 1972, 1975, 1977, 1978, 1980, 1981, 1984, 1987, 1988, 1989).

NSF National Surveys of College Graduates (NSCG)

Statistics conditioned on the actual year of college graduation are constructed from the 1993 and 2003 NSF National Surveys of College Graduates. These are representative samples of college graduates drawn from 1990 and 2000 census respondents. The 1993 and 2003 surveys of individuals who indicated in the preceding census that they were college graduates collected retrospective information about all college degrees, including field of degree and graduation dates. For my analysis, this sample is restricted to individuals born in the United States.

Project Talent

This survey of 5 percent of all 1960 U.S. high school students did not even ask students their race during the base year survey. However, the subset of original participants who were resurveyed either five or eleven years after high school graduation were asked about race and educational attainments. This sample contains fourteen black engineering graduates, first observed as high school juniors or seniors in 1960.

Current Population Surveys (CPS)

Statistics on the number of new black engineering graduates have been collected since the late 1960s, but counts of the number of new black college graduates do not begin until nearly a decade later. Estimates of the number of bachelor's degrees earned by black students in earlier years are computed based on a combination of data from different sources, including the total number of degrees conferred each year (from NCES) and estimates of the share of college graduates in the corresponding birth cohort who are black (separate estimates made from CPS data and from NSCG data are averaged).

National Science Foundation (NSF)

Additional statistics on the number of men and women graduating with degrees in engineering or computer science in the early years were drawn from National Science Foundation publications in the Surveys of Science Resources Series (National Science Foundation 1977, 1982a, 1982b, 1984; Hill 1992).

References

Abcarian, Robin. 2017. "Finally a Space to Call Their Own." *Los Angeles Times*, Feb. 22.

Alden, John D. 1970. "Engineering and Technology Degrees, 1968–69." *Engineering Education* 60:399–405.

———. 1971. "Engineering and Technology Degrees, 1969–70." *Engineering Education* 61:431–46.

Almond, Douglas, Hilary W. Hoynes, and Diane Whitmore Schanzenbach, 2011. "Inside the War on Poverty: The Impact of Food Stamps on Birth Outcomes." *Review of Economics and Statistics* 93 (2): 387–403.

Benbow, Camilla, and Olya Arjmand. 1990. "Predictors of High Academic Achievement in Mathematics and Science by Mathematically Talented Students: A Longitudinal Study." *Journal of Educational Psychology* 82 (3): 430–41.

Bix, Amy Sue. 1999. "'Engineeresses' 'Invade' Campus: Four Decades of Debate over Technical Coeducation." *Proceedings, Women and Technology: Historical, Societal, and Professional Perspectives*, 195–201. IEEE International Symposium on Technology and Society, New Brunswick, NJ, July.

———. 2004. "From Engineeresses to Girl Engineers to Good Engineers: A History of Women's American Engineering Education." *National Women's Studies Association Journal* 16 (1): 27–49.

———. 2013. *Girls Coming to Tech! A History of American Engineering Education for Women*. Cambridge, MA: MIT Press.

Blackwell, James E. 1981. *Mainstreaming Outsiders: The Production of Black Professionals*. Bayside, NY: General Hall.

———. 1987. *Mainstreaming Outsiders: The Production of Black Professionals*, 2nd ed. Dix Hill, NY: General Hall.

Blakemore, Arthur E., and Stuart A. Low. 1984. "Sex Differences in Occupational Selection: The Case of College Majors." *Review of Economics and Statistics* 66 (Feb.): 157–63.

Bound, John, and Richard B. Freeman. 1992. "What Went Wrong? The Erosion of Relative Earnings and Employment among Young Black Men in the 1980s." *Quarterly Journal of Economics* 107 (1): 201–32.

Brown, Charles, and Mary Corcoran. 1997. "Sex-Based Differences in School Content and the Male-Female Wage Gap." *Journal of Labor Economics* 15:431–65.

Card, David. 1995. "Using Geographic Variation in College Proximity to Estimate the Return to Schooling." In *Aspects of Labor Market Behaviour: Essays in Honour of John Vanderkamp*, edited by Louis N. Christofides, E. Kenneth Grant, and Robert Swidinsky. Toronto: University of Toronto Press.

Card, David, and Alan Krueger. 1992. "School Quality and Black/White Relative Earnings: A Direct Assessment." *Quarterly Journal of Economics* 107:151–200.

———. 1996. "School Resources and Student Outcomes: An Overview of the Literature and New Evidence from North and South Carolina." *Journal of Economic Perspectives* 10 (4): 31–50.

Carey, Phillip. 1977. "Engineering Education and the Black Community: A Case for Concern." *Journal of Negro Education* 46 (1): 39–45.

Chay, Kenneth Y., and Michael Greenstone. 2000. "The Convergence in Black-White Infant Mortality Rates during the 1960s." *American Economic Review* 90 (2): 326–32.

Chay, Kenneth Y., Jonathan Guryan, and Bhashkar Mazumder. 2009. "Birth Cohort and the Black-White Achievement Gap: The Roles of Access and Health Soon after Birth." NBER Working Paper no. 15078, Cambridge, MA.

———. 2014. "Early Life Environment and Racial Inequality in Education and Earnings in the United States." NBER Working Paper no. 20539, Cambridge, MA.

Chubin, Daryl, Gary S. May, and Eleanor L. Babco. 2013. "Diversifying the Engineering Workforce." *Journal of Engineering Education* 94 (1): 73–86.

College of Lewis and Clark. 2012. "Professor Emerita Awarded for Civil Service." July 10. Accessed Mar. 23, 2015. http://graduate.lclark.edu/live/news/17034 -professor-emerita-awarded-for-civic-service.

Conrad, Cecilia. 2006. "African Americans and High Tech Jobs: Trends and Disparities in 25 Cities." Joint Center for Political and Economic Studies, Washington, D.C.

Constantine, J. 1995. "The Effect of Attending Historically Black Colleges and Universities on Future Wages of Black Students." *Industrial and Labor Relations Review* 48 (3): 531–46.

Currie, Janet, and Enrico Moretti. 2008. "Did the Introduction of Food Stamps Affect Birth Outcomes in California?" In *Making Americans Healthier: Social and Economic Policy as Health Policy*, edited by R. Schoeni, J. House, G. Kaplan, and H. Pollack. New York: Russell Sage Press.

Downing, Lewis K. 1935. "The Negro in the Professions of Engineering and Architecture." *Journal of Negro Education* 4 (1): 60–70.

Ehrenberg, Ronald G., and Donna S. Rothstein. 1994. "Do Historically Black Institutions of Higher Education Confer Unique Advantages on Black Students?" In *Choices and Consequences: Contemporary Policy Issues in Education*, edited by Ronald Ehrenberg. Ithaca, NY: ILR Press.

Engineering Manpower Commission (EMC). 1972. *Engineering and Technology Degrees, 1972*. New York: Engineers Joint Council.

———. 1975. *Engineering and Technology Degrees, 1974*. New York: Engineers Joint Council.

———. 1977. *Engineering and Technology Degrees, 1976*. New York: Engineers Joint Council.

———. 1978. *Engineering and Technology Degrees, 1978*. New York: Engineers Joint Council.

———. 1980. *Engineering and Technology Degrees, 1980*. New York: American Association of Engineering Societies.

———. 1981. *Engineering and Technology Degrees, 1981*. New York: American Association of Engineering Societies.

———. 1984. *Engineering and Technology Degrees, 1984*. New York: American Association of Engineering Societies.

———. 1987. *Engineering and Technology Degrees, 1986*. New York: American Association of Engineering Societies.

———. 1988. *Engineering and Technology Degrees, 1987*. New York: American Association of Engineering Societies.

———. 1989. *Engineering and Technology Degrees, 1988*. New York: American Association of Engineering Societies.

Fiorito, Jack, and Robert Dauffenbach. 1982. "Market and Nonmarket Influences on Curriculum Choice by College Students." *Industrial and Labor Relations Review* 36 (1): 88–101.

Freeman, Richard B. 1976a. *Black Elite: The New Market for Highly Educated Black Americans: A Report Prepared for the Carnegie Commission on Higher Education*. New York: McGraw-Hill.

———. 1976b. "A Cobweb Model of the Supply and Starting Salary of New Engineers." *Industrial and Labor Relations Review* 29 (2): 236–48.

Fries-Britt, Sharon, Brian A. Burt, and Khadish Franklin. 2012. "Establishing Critical Relationships, How Black Males Persist in Physics at Historically Black Colleges and Universities." In *Black Men in College: Implications for HBCUs and Beyond*, edited by Robert T. Palmer and J. Luke Wood, 71–88. New York: Routledge.

Fryer, Roland, and Michael Greenstone. 2010. "The Changing Consequences of Attending Historically Black Colleges and Universities." *American Economic Journal: Applied Economics* 2 (1): 116–48.

Garces, Eliana, Duncan Thomas, and Janet Currie. 2002. "Longer-Term Effects of Head Start." *American Economic Review* 92 (4): 999–1012.

Goldin, Claudia, and Lawrence F. Katz. 1999. "The Shaping of Higher Education: The Formative Years in the United States, 1890–1940." *Journal of Economic Perspectives* 13 (1): 37–62.

Goodman, Joshua. 2017. "The Labor of Division: Returns to Compulsory High School Math Coursework." NBER Working Paper no. 23063, Cambridge, MA.

Grogger, Jeff, and Eric Eide. 1995. "Changes in College Skills and the Rise in the College Wage Premium." *Journal of Human Resources* 30 (2): 280–310.

Hill, Susan. 1992. *Science and Engineering Degrees, by Race/Ethnicity of Recipients, 1977–90: Detailed Statistical Tables*. Surveys of Science Resources Series. Washington, D.C.: National Science Foundation.

Hoynes, Hilary W., Diane Whitmore Schanzenbach, and Douglas Almond. 2016. "Long Run Impacts of Childhood Access to the Safety Net." *American Economic Review* 106 (4): 903–34.

Humphrey, Senator Hubert. 1973. "A Tenfold Increase in Minority Engineers—A

Civil Rights Challenge for the Seventies." Recorded in U.S. Congressional Record: Proceedings and Debates of the 93rd Congress, Jan. 12, 1973, 1040–43. 93rd Cong., 1st sess., v. 119: pt. 34 (1973: Jan. 3/Dec. 22). Washington, D.C.: Superintendent of Documents, U.S. Government Printing Office.

IBM. 1973. "When Bob Lee went to Tuskegee Institute as Part of IBM's Faculty Loan Program, the Students Expected Him to Teach Electronics. They Got a Lot More Than They Expected." Paid advertisement, *Black Enterprise*, March.

Johnson, Rucker C. 2010. "The Health Returns of Education Policies from Preschool to High School and Beyond." *American Economic Review* 100 (2): 188–94.

———. 2011. "Long-Run Impacts of School Desegregation & School Quality on Adult Attainments." NBER Working Paper no. 16664, Cambridge, MA.

Kane, Thomas. 1998. "Racial and Ethnic Preferences in College Admissions." In *The Black-White Test Score Gap*, edited by Christopher Jencks and Meredith Phillips. Washington, D.C.: Brookings Institution Press.

Landis, Raymond B. 2005. *Retention by Design, Achieving Excellence in Minority Engineering Education*. College of Engineering, Computer Science and Technology, California State University, Los Angeles. https://ai2-s2-pdfs.s3.amazonaws .com/83f1/7c12bedfd483033576e744be5b3bc0db70ab.pdf.

Lusterman, Seymour. 1979. *Minorities in Engineering: The Corporate Role*. New York: The Conference Board.

Miranda, Luis, and Esther Ruiz. 1986. *NACME Statistical Report, 1986*. New York: National Action Council for Minorities in Engineering. https://eric.ed.gov /?id=ED274761.

Murnane, Richard J., John B. Willet, and Frank Levy. 1995. "The Growing Importance of Cognitive Skills in Wage Determination." *Review of Economics and Statistics* 77 (May): 251–66.

Mykerezi, Elton, and Bradford F. Mills. 2008. "The Wage Earnings Impact of Historically Black Colleges and Universities." *Southern Economic Journal* 75 (1): 173–87.

Mykerezi, Elton, Bradford F. Mills, and Sonya Gomes. 2003. "Education and Socioeconomic Well-Being in Racially Diverse Rural Counties." *Journal of Agricultural and Applied Economics* 35:251–62.

NACME. 1986. *Student's Guide to Engineering Schools*. New York: National Action Council for Minorities in Engineering. https://eric.ed.gov/?id=ED276554.

National Academy of Engineering & Symposium on Increasing Minority Participation in Engineering (NAE). 1973. Proceedings of Symposium on Increasing Minority Participation in Engineering, Washington, D.C., May 6–8.

National Center for Education Statistics (NCES). 1994. *NAEP Trends in Academic Progress*. Washington, D.C.: U.S. Government Printing Office. (See pages 80, 83, and A-69 for discussion of six National Assessment of Educational Progress Mathematics Trend Assessments spanning 1973–1992). https://archive.org /details/ERIC_ED378237.

National Science Foundation. 1977. *Women and Minorities in Science and Engineering*. Surveys of Science Resources Series NSF 77-304. Washington, D.C.: National Science Foundation.

———. 1982a. *Science and Engineering Degrees 1950–1980: A Source Book*. Surveys of Science Resources Series NSF 82-307. Washington, D.C.: National Science Foundation.

———. 1982b. *Women and Minorities in Science and Engineering*. Surveys of Science Resources Series NSF 82-302. Washington, D.C.: National Science Foundation.

———. 1984. *Women and Minorities in Science and Engineering*. Surveys of Science Resources Series NSF 84-300. Washington, D.C.: National Science Foundation.

Nettles, Michael. 1988. *Toward Black Undergraduate Student Equality in American Higher Education.* Westport, CT: Greenwood Press.

Padulo, Louis. 2015. "Louis Padulo—Biographical Sketch." Accessed Mar. 20, 2015. http://www.padulo.org/bio.html.

Paglin, Morton, and Anthony M. Rufolo. 1990. "Heterogeneous Human Capital, Occupational Choice, and Male-Female Earnings Differences." *Journal of Labor Economics* 8 (Jan.): 123–44.

Pierre, Percy. 1972. "Remarks by Dr. Percy Pierre, Dean of Engineering, Howard University." In *Minorities in Engineering, Proceedings of a Meeting of the Engineering Manpower Commission of Engineers Joint Council, November 30, 1972.* New York: Engineers Joint Council.

———. 1975. "Keynote Address." In *Proceedings of a Workshop for Program Directors in Engineering Education of Minorities, Conducted by the Committee on Minorities in Engineering, Assembly of Engineering,* 55–57. Washington, D.C.: National Academy of Sciences National Research Council. https://eric.ed.gov /?id=ED149882.

———. 2012. Personal interview. Feb. 29.

———. 2013. Personal correspondence, Apr. 12. (Outline of a chapter in progress on the history of the Minority Engineering Education Effort, untitled.)

———. 2015. "A Brief History of the Collaborative Minority Engineering Effort." In *Changing the Face of Engineering: The African American Experience,* edited by John Brooks Slaughter, Yu Tao, and Willie Pearson Jr. Baltimore: Johns Hopkins University Press.

Planning Commission for Expanding Minority Opportunities in Engineering (PCEMOE). 1974. *Minorities in Engineering: A Blueprint for Action: A Report by the Planning Commission for Expanding Minority Opportunities in Engineering.* New York: Alfred P. Sloan Foundation.

Pruitt, Anne S. 1987. *In Pursuit of Equality in Higher Education.* Dix Hills, NY: General Hall.

Royster, Deirdre A. 2003. *Race and the Invisible Hand: How White Networks Exclude Black Men from Blue-Collar Jobs.* Berkeley: University of California Press.

Ruggles, Steven, J. Trent Alexander, Katie Genadek, Ronald Goeken, Matthew B. Schroeder, and Matthew Sobek. 2010. Integrated Public Use Microdata Series: Version 5.0 [Machine-readable database]. Minneapolis: University of Minnesota.

Saline, Lindon E. 1974. "A National Effort to Increase Minority Engineering Graduates." *IEEE Transactions on Education* 17 (1): 1–73.

Schneller, Robert J. 2008. *Blue & Gold and Black: Racial Integration of the U.S. Naval Academy.* College Station: Texas A&M University Press.

Sibulkin, Amy E., and J. S. Butler. 2005. "Differences in Graduation Rates between Young Black and White College Students: Effect of Entry into Parenthood and Historically Black Universities." *Research in Higher Education* 46 (3): 327–48.

Slaughter, John Brooks. 2009. "African American Males in Engineering: Past, Present and a Future of Opportunity." In *Black American Males in Higher Education: Research, Programs and Academe,* Diversity in Higher Education 7, edited by Henry Frierson, James Wyche, and Willie Pearson, 193–208. Bingley, U.K.: Emerald Publishing.

———. 2012. Personal interview, Feb. 24.

Smith, J. Stanford. 1973. "Needed: A Tenfold Increase in Minority Engineering Graduates." Recorded in U.S. Congressional Record: Proceedings and Debates of the 93rd Congress, Jan. 12, 1973, 1041–43. 93rd Cong., 1st sess., v. 119: pt. 34 (1973: Jan. 3/Dec. 22). Washington, D.C.: Superintendent of Documents, U.S. Government Printing Office.

Tolliver, Wayne E., and Henry H. Armsby. 1959. *Engineering Enrollments and Degrees, 1959*. Washington, D.C., U.S. Dept. of Health, Education, and Welfare, Office of Education.

———. 1961. *Engineering Enrollments and Degrees: 1960*. Washington, D.C.: U.S. Dept. of Health, Education, and Welfare, Office of Education.

———. 1963. *Engineering Enrollments and Degrees, 1961*. Washington, D.C.: U.S. Government Printing Office.

Treisman, Uri. 1992. "Studying Students Studying Calculus: A Look at the Lives of Minority Mathematics Students in College." *College Mathematics Journal* 23 (5): 362–72.

Trent, William, and John Hill. 1994. "Contributions of Historically Black Colleges and Universities to the Production of African American Scientists and Engineers." In *Who Will Do Science? Educating the Next Generation*, edited by Willie Pearson Jr. and Alan Fechter, 68–80. Baltimore: Johns Hopkins University Press.

Turner, Sarah, and William Bowen. 1999. "Choice of Major: The Changing (Unchanging) Gender Gap." *Industrial and Labor Relations Review* 52 (2): 289–313.

U.S. Office of Education. 1942/1943. *National Survey of the Higher Education of Negroes*. Washington, D.C.: Government Printing Office.

Weinberg, Meyer. 1977. *A Chance to Learn, the History of Race and Education in the United States*. Cambridge: Cambridge University Press.

Weinberger, Catherine J. 1998. "Race and Gender Wage Gaps in the Market for Recent College Graduates." *Industrial Relations* 37 (1): 67–84.

———. 1999. "Mathematical College Majors and the Gender Gap in Wages." *Industrial Relations* 38 (3): 407–13.

———. 2001. "Is Teaching More Girls More Math the Key to Higher Wages?" In *Squaring Up: Policy Strategies to Raise Women's Incomes in the U.S.*, edited by Mary C. King, 50–72. Ann Arbor: University of Michigan Press.

———. 2014. "Are There Racial Gaps in High School Leadership Opportunities? Do Academics Matter More?" *Review of Black Political Economy* 41 (4): 393–409.

Weinberger, Catherine J., and Lois Joy. 2007. "The Relative Earnings of Black College Graduates, 1980–2001." In *Race and Economic Opportunity in the 21st Century*, edited by Marlene Kim, 226–55. London: Routledge.

Wharton, D. E. 1992. *A Struggle Worthy of Note: The Engineering and Technological Education of Black Americans*. Westport, CT: Greenwood Press.

Williams, Clarence. 1999. *Technology and the Dream: Reflections on the Black Experience at MIT, 1941–1999*. Cambridge, MA: MIT Press.

Witherell, Carol. 2009. *Quiet Water Seen through Trees, the Life of J. Stanford Smith*. South Barre, VT: Acura Printing.

Xie, Yu, and Kimberlee A. Shauman. 2003. *Women in Science: Career Processes and Outcomes*. Cambridge, MA: Harvard University Press.

4

Bridging the Gaps between Engineering Education and Practice

Samantha R. Brunhaver, Russell F. Korte,
Stephen R. Barley, and Sheri D. Sheppard

Increasingly, American engineers contend with challenges at work including rapid technological innovation and the needs of changing workplaces (Duderstadt 2008; National Academy of Engineering 2008b; National Research Council 2007). In response, industry, government, and professional societies have called on educators to better prepare engineering students by emphasizing not only technical but professional competencies (Jamieson and Lohmann 2009; Sheppard et al. 2008; Shuman, Besterfield-Sacre, and McGourty 2005). There is a consensus in the engineering community that those competencies include communication skills, business skills, teamwork skills, creativity, lifelong-learning skills, and problem-solving skills (ABET 2011; American Society of Civil Engineers 2008; McMasters and Matsch 1996; National Academy of Engineering 2004).

Yet, despite calls for reform, engineering programs are often based on an outdated image of engineering practice that is misaligned with reality (Duderstadt 2008; National Research Council 2007; Sheppard et al. 2008;

Samantha R. Brunhaver is assistant professor of engineering at The Polytechnic School at Arizona State University. Russell F. Korte is associate professor in Organizational Learning, Performance, and Change at Colorado State University. Stephen R. Barley is the Christian A. Felipe Professor of Technology Management at the College of Engineering at the University of California, Santa Barbara. Sheri D. Sheppard is the Burton J. and Deedee McMurtry University Fellow in Undergraduate Education and professor of mechanical engineering at Stanford University.

This research was supported by two National Science Foundation (NSF) grants no. 0227558 and no. 1022644, as well as a Stanford Graduate Fellowship from Stanford University. The authors thank Michelle Grau and Michelle Warner for their assistance with coding, the editors for helpful comments on earlier versions of this chapter, and the engineers who participated in our research. For acknowledgments, sources of research support, and disclosure of the authors' material financial relationships, if any, please see http://www.nber.org/chapters /c12687.ack.

Vest 2008). Although society has long looked to higher education to develop the nation's workforce (Sullivan and Rosin 2008), employers, faculty, and even students have questioned whether engineering programs meet this goal. For example, in 2004 almost a quarter of employers reported that engineering graduates were less skilled in problem solving and less aware of organizational contexts and constraints than graduates ten years earlier. A third of the teaching faculty reported that the students were less able in math and science and had worse technical skills than their forerunners (Lattuca, Terenzini, and Volkwein 2006). Other research has shown that many engineering students remain uncertain about what engineering is and what engineers do, even by the time they graduate (Matusovich et al. 2009).

Our interviews together with other studies identify shortcomings that engineering programs have in preparing students for practice. First, as more and more U.S. corporations downsize and outsource, graduates looking for engineering work face pressure to stay relevant and adapt to new industry needs (Shuman, Besterfield-Sacre, and McGourty 2005). Second, graduates may need additional training to acquire missing competencies, which comes at a cost to their employers (Salzman 2007). Third, uncertainty about what engineering really involves and feelings of being ill prepared may convince some graduates to abandon engineering. Two years after graduation 28 percent of all engineering graduates from the classes of 2008 and 2009 were working in fields outside of engineering, including mathematics and computer science (National Science Foundation 2010). Similarly, 20 percent of engineering seniors in 2007 reported that they were unsure about pursuing an engineering career (Atman et al. 2010). Their uncertainty partially reflects a nationwide loss of interest in engineering, a belief that engineering careers are no longer secure, and the realization that other professions pay higher wages (Chubin, May, and Babco 2005; Duderstadt 2008; Lowell and Salzman 2007; Shuman, Besterfield-Sacre, and McGourty 2005). Students' limited understanding of what engineering entails and the skills it requires may also lead them to reject engineering careers (Lichtenstein et al. 2009; National Academy of Engineering 2008a).

The studies described above support Trevelyan's (2007) claim that a better understanding of engineering practice is needed to inform students' career choices and improve their preparedness to work in this field. One of the best ways to accomplish this is to study engineering students and young engineers who have just entered the workforce and the challenges they encounter and how they make sense of these challenges. This chapter uses interviews with engineering students and newly hired engineers from the Academic Pathways Study (Sheppard et al. 2009) and the Engineering Pathways Study (Sheppard et al. 2011) to examine disjunctures between engineering education and professional practice and describes possible ways to better connect training to practice.

4.1   Background

Although most studies of how to prepare engineering students more effectively for professional practice focus on mastering competencies that educators and employers think are important (e.g., Bankel et al. 2003; Lattuca, Terenzini, and Volkwein 2006; Meier, Williams, and Humphreys 2000), some studies approach the problem by exploring how young engineers perceive the relevance of their education. Most of the latter have studied young engineers in the workplace, while a few have studied engineering students currently in school. Dunsmore, Turns, and Yellin (2011) found that mechanical engineering students sharply contrasted their academic experience with professional practice. Students thought of work, but not school, as the "real world." They identified teamwork and communication as real-world skills crucial for success, but still believed that such skills were less important than skills like the ability to design. Seniors surveyed in the Academic Pathways Study also said communication and teamwork were important skills (Atman et al. 2010), but thought their school experiences were unrelated to engineering work (Sheppard et al. 2010). Even the 50 percent who reported exposure to engineering through internships, jobs, company visits, and research experiences had no greater appreciation for the relevance of professional and interpersonal skills than did freshmen (Sheppard et al. 2010; Atman et al. 2010). These results suggest that while senior engineering students think that communication and teamwork are important, they may not understand how these skills are actually used in practice.

Many more studies have asked practicing engineers how well their undergraduate education prepared them for their current work. Across these studies, working engineers report that knowing how to communicate and work with other people is paramount and that most of what they now do was learned on the job. Roughly half of MIT's mechanical engineering alumni from the classes of 1992–1996 reported using professional skills,[1] interpersonal skills, and independent-thinking skills almost daily in their work and said that they had learned these skills on the job (Seering 2009). In contrast, the MIT alumni reported hardly ever using the theoretical knowledge they learned in college (for example, thermodynamics and fluid mechanics). Almost identical results were found in a study of practicing engineers as part of the National Science Foundation (NSF)-funded Aligning Educa-

---

1. Seering (2009) operationalized professional skills as professional ethics and integrity, responsibility and accountability, professional behavior, proactively planning for one's career, and continuous learning; personal skills as initiative and willingness to take risks, perseverance and flexibility, creative thinking, and time and resource management; and independent thinking skills as skills in working independently, skills in setting project goals, ability to extract and evaluate relevant knowledge; and confidence in one's own skills and abilities.

tional Experiences with Ways of Knowing Engineering (AWAKEN) project. The respondents in this study considered communication, problem solving, ethics, lifelong learning, and business skills to be essential to their work (Anderson et al. 2009). Two-thirds said that their schooling prepared them well for engineering practice and half said they still used skills they learned as undergraduates. But for many, "real" engineering consisted only of technical problem solving (Grohowski-Nicometo, Nathans-Kelly, and Anderson 2009), and most said that beyond the ability to think analytically, what they need to know at work they learned after graduating (Anderson et al. 2011). Passow (2012) found that engineers who graduated from a large, midwestern university in the 1990s and first decade of the twenty-first century also rated teamwork, communication, problem solving, and analytical skills as important in their professional experience. As a part of the Prototyping the Engineer of 2020 (P2P) study, Lattuca and colleagues (Lattuca et al. 2014) discovered that engineering alumni from thirty-one institutions rated communication skills, teamwork skills, and professional skills as highly important to their current work. Of these three, teamwork was perceived as having been the most heavily emphasized in their undergraduate curricula.[2]

Finally, as part of the NSF-funded Academic Pathways Study, Korte, Sheppard, and Jordan (2008) interviewed young engineers at a global automobile manufacturer to identify differences between their experiences at work and school and found that the company's social and organizational context set the problems and processes on which the engineers worked. These problems were often more complex, ambiguous, and political than those the engineers had encountered in school. Additionally, how the engineers perceived and learned about engineering work depended to a large extent on their interactions with coworkers in their work groups (Korte 2009, 2010).

The foregoing studies illuminate the relation between engineering education and practice, but have limitations that this chapter seeks to fill. More than half of the studies rely solely on survey data (Anderson et al. 2009; Atman et al. 2010; Lattuca et al. 2014; Passow 2012; Seering 2009; Sheppard et al. 2010; Trevelyan and Tilli 2008; Trevelyan 2008) that allow participants to voice their perspectives within the scope of defined response options, but

---

2. American engineers are not unique in their perceptions of their education's relevance. In a longitudinal study of early career professionals in the United Kingdom, most engineers differentiated between learning in school, which emphasized theory, and learning in the workplace, which emphasized communication, teamwork, leadership, decision making, reflection, and awareness (Eraut 2009). In another longitudinal study, Trevelyan and Tilli (2008) found engineers who graduated from the University of Western Australia (UWA) in 2006 spent, on average, 60 percent of their time interacting with other people at work. Ten percent said they wished they had been taught interpersonal skills at UWA, and interpersonal skills was the area in which engineers were most likely to report needing further training. Like American engineers, the Australians reported learning most of what they needed know on their own or from their coworkers (Trevelyan 2008).

that give no opportunity for researchers to acquire a deeper contextually sensitive understanding of young engineers' experiences at work. Most of the studies sample engineers from or working in a handful of schools or organizations.[3] Most important, each study captured only one side of the school-to-work transition, asking engineering students to anticipate future engineering practice, or asking engineering practitioners to reflect back on their academic preparation.[4] To add to the existing knowledge, this chapter uses information obtained by structured and semistructured interviews that elicit participants' own perspectives on the relationship between their engineering education and practice. It uses two samples from the Academic Pathways Study (APS):[5] a *Workplace Sample* of newly hired engineers and a *Longitudinal Sample* of junior-year engineering students in the APS who were later interviewed as early career professionals in the follow-on Engineering Pathways Study (EPS) (see Sheppard et al. 2011).

The Workplace Sample comprises fifty-seven newly hired engineers in their first full-time position since graduating, with whom we conducted semistructured forty-five- to ninety-minute interviews in the winter of 2007. As recent graduates, these individuals could best offer opinions about the differences between their engineering education and new employment. The participants worked in four U.S. organizations: a global automobile manufacturer (Car Company), a large food company (Food Manufacturer), a smaller computer components manufacturer (Computer Parts Company), and a state transportation agency (Transportation Department). The engineers at Car Company were mostly mechanical engineers, those at Food Manufacturer were chemical engineers, those at the Computer Parts Company were mechanical and chemical engineers, and those at Transportation Department were civil engineers. Car Company and Computer Parts Company assigned young engineers to permanent positions that involved both technical and project management work. Young engineers at Food Manufacturer and Transportation Department initially rotated through at least three departments before being assigned to a project management position.

To identify what young, practicing engineers need to know and whether they had learned that knowledge in school or on the job, we asked participants to reflect on one or two problems or projects to which they had been assigned and asked: "What knowledge and skills did you apply to work on the problem?" and "Where and how did you learn the knowledge and skills?" We also asked participants to reflect on their socialization experiences since

3. The exceptions are the Academic Pathways Study, the P2P study, and the AWAKEN study.
4. Even the two longitudinal studies (Eraut 2009; Trevelyan and Tilli 2008) focused exclusively on recently graduated engineers.
5. Funded by NSF, the APS was conducted between 2003 and 2007 under the auspices of the Center for the Advancement of Engineering Education (CAEE). Atman et al. (2010) provides details of the study's design and an overview of its findings.

joining their organization and asked: "How did you learn the way things work at this company?" and "How did you learn what others expect of you on the job?" Finally, we asked participants about differences between their expectations as engineering students in school and experiences on the job.

Throughout the interviews, participants talked readily about knowledge and skills they used in their assignments and their work in general, differences between school and work, how much they used their overall education, and how well their education prepared them for their jobs. Passages in the interviews that deal with these topics, as well as the participants' answers to the questions above, provided data for the present analysis.

Table 4.1 summarizes demographic information on the Workplace Sample. All participants had been in their jobs for two months to three years, with an average duration of eleven months. Forty-five had internships in college and eighteen had interned with their current employer. Nine participants had master's degrees, and the rest had bachelor's degrees. Sixteen of the fifty-seven worked at Car Company, eighteen at Food Manufacturer, sixteen at Computer Parts Company, and seven at Transportation Department. Codes describing the knowledge and skills respondents used in completing specific assignments, learning about their organization, and their work in general emerged inductively from the data analysis and stayed close to the participants' own language (Patton 2002).

The Longitudinal Sample comprises two sets of interviews from three institutions in the western United States: a Technical Public institution (TPub), a Suburban Private university (SPri), and a Large Public university (LPub). The first set of interviews was conducted under the Academic Pathways Study (APS) in 2006, when the participants were junior-year engineering students (Sheppard et al. 2009). The second set of interviews were part of the Engineering Pathways Study (EPS) in 2011, four years after the participants earned their engineering bachelor's degrees (Sheppard et al. 2011). The two interviews allow us to compare how the same individuals thought about engineering as students with how they understood engineer-

Table 4.1   Demographic information for participants in the Workplace Sample ($n = 57$)

	Count	Average time since hire (months)	Number who reported internship		Number who reported earning master's degree
			At any company	At current company	
All participants	**57**	**11.3**	**45**	**18**	**9**
Food Manufacturer	18	13.7	15	3	1
Car Company	16	11.4	14	8	7
Computer Parts Company	16	7.0	9	0	1
Transportation Department	7	15.4	7	7	0

Table 4.2   **Demographic information for participants in the Longitudinal Sample ($n = 9$)**

Participant	Institution	Sex	URM	Major reported in junior year	Reported internships in junior year	Occupation reported at follow-up
1	LPub	Female	Yes	Civil and environmental engineering	Yes	Structural engineer
2	LPub	Male	No	Mechanical engineering	No	General engineer
3	TPub	Male	No	Petroleum engineering	Yes	Operations engineer
4	LPub	Male	No	Computer engineering	No	Software engineer
5	TPub	Female	No	Chemical engineering	n/a	Production engineer
6	TPub	Female	No	Chemical engineering	Yes	Process engineer
7	TPub	Female	Yes	Chemical engineering	No	Process engineer
8	SPri	Male	No	Computer science	Yes	User interface engineer
9	TPub	Male	Yes	Engineering physics	No	R&D engineer

Notes: LPub = Large Public University, SPri = Suburban Private University, TPub = Technical Public Institution, URM = Underrepresented minority (e.g., not white or Asian/Asian American), n/a = information not available.

ing after four years work experience. The sample consisted of nine engineers working in an engineering job at the time of the second interview: five from TPub, three from LPub, and one from SPri. The small sample notwithstanding, these interviews offer insight on how engineers' perceptions change over time, which we then seek to check with our Workplace Sample.

Table 4.2 displays demographic information for the Longitudinal Sample participants. Four of the nine were women, and four were from an underrepresented minority. Three were chemical engineering majors, two were computer-related majors, and the remainder ranged across other engineering disciplines. Four had held internships by their junior year. At the time of the interviews, one was enrolled in an engineering master's program and the other eight were working. Researchers interviewed the participants in person for approximately an hour when they were college juniors and on the phone for thirty to sixty minutes five years later. We examine first participants' answers to three questions asked when they were juniors in the spring of 2006:

- Are there particular skills that you would say are important for an engineer to have?
- Of the skills you mentioned, which ones do you possess and how did you develop them?
- In your own words, please define engineering.

We then focus on their answers to two questions posed during our follow-up interviews in spring 2011:

- What knowledge and skills do you see as most important to doing your job?
- How has your idea of an engineering job changed since you graduated?

4.2   Methodology of Interview Analysis

We transcribed each set of interviews verbatim and made them anonymous. We then analyzed the interviews for each sample following a case study approach, with each participant representing a case (Creswell 1998; Miles and Huberman 1994; Stake 2006; Yin 2003). We used MAXQDA software to code the data. In this chapter, we use retrospective data from the Workplace Sample to identify how the knowledge and skills employed by young engineers diverged from what they learned in school. We combine the longitudinal data from the APS and the EPS to show how work altered the ideas about engineering held by participants as undergraduates. We then return to the Workplace Sample for further evidence on how young engineers' images of engineering changed since leaving school.

In the Workplace Sample, after coding each transcript, we combined mentions of different knowledge and skills into twenty-three distinct categories such as content knowledge and communication skills. We then clustered the categories into three broad types: technical knowledge and skills, professional knowledge and skills, and organizational knowledge and skills. In our analysis, *knowledge* relates to understanding or awareness of concepts, principles, and information related to a specific domain and *skill* to the ability to apply domain knowledge in a particular context. For instance, an engineer might have technical knowledge of how a machine works, but would use her technical skills to troubleshoot it. *Technical knowledge and skills* refers to competencies to accomplish specific engineering, mathematical, scientific, or computer-related tasks, such as technical problem solving, analysis, and design. In contrast, *professional knowledge and skills* are nontechnical competencies that relate to the profession of engineering and interaction with people more generally (Jarosz and Busch-Vishniac 2006, 243). Sometimes referred to as "soft" or "social" skills, professional knowledge and skills include teamwork, communication, and leadership skills (ABET 2011; Knight 2012; Shuman, Besterfield-Sacre, and McGourty 2005). Finally, *organizational knowledge and skills* relate to one's organization, work group, and job role that enables one to better navigate the work (Korte, Sheppard, and Jordan 2008; Korte 2009, 2010).[6]

We quantified the data using frequency counts and tables to describe the sample and identify patterns (Sandelowski, Voils, and Knafl 2009). For each skill and area of knowledge, we counted the percentage of participants who

6. See the appendix for definitions for each skill and area of knowledge.

reported using it, the percentage of participants who reported learning it in school, and the percentage of participants who reported learning it (or continuing to learn it) on the job. We coded differences in the Workplace Sample's perceptions of school and work in the same way that we coded knowledge and skills, but we present these results in a more qualitative fashion.

In the Longitudinal Sample, we coded the interviews for the knowledge and skills respondents considered important to doing engineering work and for their ideas about what engineering entails. The longitudinal interviews were analyzed using the same inductive coding used for the Workplace Sample, but differ slightly since they are grounded in the language our informants used. Furthermore, because the Longitudinal Sample spans two sets of interviews, a consistent coding scheme was required across sets. For example, we grouped mentions of variations on problem-solving skills when participants were juniors and when they were four years beyond their bachelor's degrees under a single category. Skills and areas of knowledge mentioned only when participants were either juniors or young engineers are represented by their own categories. Each skill and area of knowledge is shown in the findings section and defined in the appendix. We also present qualitative results related to participants' changing perceptions of engineering work.

The two samples complement each other in giving the insights they provide into the "gaps" between engineering education and practice. The Workplace Sample of young engineers working for zero to three years in their first jobs after graduation helped us compare knowledge and skills engineers use at work with those learned in school. The Longitudinal Sample of early career engineers as college juniors and then as practitioners four years out offers a unique picture of changes in engineers' views of the work they do. Table 4.3 summarizes the way the surveys fit together in our analysis.

Table 4.3   Summary of Workplace and Longitudinal Samples

Sample	Data source	Sample description	Sample scope	Main focus of study
Workplace Sample	APS workplace cohort	Fifty-seven recently graduated engineers, zero to three years out in 2007	Four organizations	Where knowledge and skills used at work are learned (in school/on the job)
Longitudinal Sample	APS/EPS longitudinal cohort	Nine early career engineers: junior college students in 2006 and graduates four years out in 2011	Three institutions	What knowledge and skills are most important to doing engineering job

Note: APS = Academic Pathways Study; EPS = Engineering Pathways Study.

4.3   Findings

4.3.1   Knowledge and Skills in Engineering Practice
(the Workplace Sample)

The recently graduated engineers in the Workplace Sample identified the knowledge and skills that they applied on the job. Table 4.4 shows the number of participants who mentioned using technical knowledge and skills, professional knowledge and skills, and organizational knowledge and skills (defined in the appendix), as well as whether each of these was learned in

Table 4.4   **Comparison of where knowledge and skills learned (Workplace Sample, $n = 57$)**

	Percentage who reported using skill/knowledge at work[a]	Percentage who reported learning skill/knowledge	
		In school	On the job
Technical knowledge and skills[b]	**98**	**93**	**86**
Content knowledge	72	61	16
Equipment/process knowledge	60	5	60
Problem-solving skills	46	44	7
Software skills	39	33	26
Modeling and analysis skills	30	25	11
Rapid iteration skills	18	11	9
Programming skills	11	9	7
Design skills	11	7	4
Testing skills	11	4	7
Systems-thinking skills	5	5	0
Hands-on skills	4	4	2
Professional skills	**96**	**53**	**96**
Communication skills	65	25	58
Working with people	60	19	49
Information-finding skills	58	16	51
Project management skills	28	18	12
Time management skills	18	5	14
Leadership skills	12	5	7
Documentation skills	11	0	11
Context knowledge	9	0	9
Work ethic	5	0	5
Organizational knowledge and skills	**67**	**16**	**56**
Organizational policies and procedures	42	7	35
Organizational hierarchy and structure	28	2	28
Organizational background and culture	16	7	9

[a]All percentages are based on the total number of participants, $n = 57$.

[b]Percentage of participants who reported using/learning at least one of the technical knowledge and skills listed in the table. The percentage of participants who reported using/learning professional knowledge and skills and the percentage of participants who reported using/learning organizational knowledge can be interpreted similarly.

school and/or on the job. As the table shows, virtually every participant reported using technical (98 percent) and professional (96 percent) knowledge and skills in their work, and most reported learning *technical knowledge and skills* in school (93 percent) and in the workplace (86 percent). But where they learned *professional knowledge and skills* differed. Just over half reported learning a professional skill or area of knowledge in school. Yet every engineer who mentioned professional knowledge and skills claimed to have mastered at least one of them at work. Additionally, only two-thirds reported relying on *knowledge about their organization* to do their work, which was learned almost entirely on the job (56 percent compared to 16 percent who reported exposure to this knowledge in school).

Although participants claimed to have acquired technical knowledge and skills from school as well as work, school was the more frequently mentioned source. Nearly three-quarters of the interviewees reported using content knowledge in their jobs, and 60 percent traced this knowledge back to courses ranging from basic physics and calculus to more advanced engineering topics (table 4.4). The advanced courses the participants viewed as relevant varied by discipline and by the companies for which they worked. Mechanical engineers at Car Company spoke of using what they learned in their engine courses, while their electrical engineering colleagues at Car Company relied more on their training in circuits. By contrast, only 16 percent of participants reported instances of content knowledge learned at work, typically around the engineer's particular job (e.g., traffic signal guidelines at Transportation Department; mechanical behavior of specific materials at Computer Parts Company).

School was also more important for learning how to solve problems and model and analyze data, with more than twice the number of participants reporting learning these skills in school than at work. With regard to problem solving, respondents felt their entire engineering education contributed to their ability to solve problems rather than any particular course or combination of courses:

> There's definitely not a class on that [problem solving] by any means. It's more as you go through problems for homework and you work with your team members in labs, you just learn how to approach problems and come up with a solution. I wouldn't say it's something that people teach you at school. It's just something you pick up along the way. (Food Manufacturer, Chemical Engineering major, seventeen months on the job)[7]

With regard to learning software applications, participants reported that school and the workplace were more or less equally important. In fact, some engineers ended up deepening their knowledge of applications that they had

---

7. To assist the reader in interpreting our qualitative data, passages from interviews end with the organization where the participant worked or where he or she went to school, their undergraduate major, and the amount of time they had been either on the job or in/out of school.

first encountered in a course. Nearly half of the engineers who learned to use a new software application on the job told us they had learned a simpler version in school, which made their learning curve less steep:

> I've used different programs up at school. It just so happens that, in this job as a process engineer, we don't deal with those ones as much, but other ones. But just having the experience of using those programs, I was able to jump right into these new ones and get a feel for them. (Computer Parts Company, Mechanical Engineering major, nine months on the job)

In contrast, young engineers overwhelmingly reported learning about specific equipment and processes pertaining to their work on the job rather than at school. This result is expected given the diversity of jobs that engineers take and the limitations of engineering curricula to train for them all. The few participants who attributed their knowledge of a process to school cited learning about it in a particular course or an internship experience. Participants mentioned a variety of other technical skills, but with far less frequency: rapid iteration skills, programming skills, design skills, testing skills, systems-thinking skills, and hands-on skills. With the exception of testing skills, the young engineers told us they had learned these skills primarily or exclusively at school.

Most interviewees reported learning *professional knowledge and skills* on the job rather than in school, and most equated learning to be professional with acquiring communication skills, working with other people, and developing information-finding skills. While we present these skills as separate categories, they are highly related. Effective oral and written communication was the most commonly used professional skill cited by 65 percent of the engineers; while a quarter reported learning this skill in school, more than twice as many (58 percent) said they learned it on the job. For participants, learning to communicate effectively encompassed not only delivering better reports and presentations, but learning how to communicate with people who were not engineers (e.g., marketing, purchasing, sales, and client personnel).

Perhaps because three of the four employers were manufacturers, many participants reported needing to communicate with—and sometimes even manage—workers on the production floor, including foremen, machine operators, and maintenance workers. Others found they needed to communicate with contractors, suppliers, and customers. Learning to work with members of such groups was an eye-opening experience for many young engineers in which they discovered the importance of expressing themselves clearly, listening carefully, and getting to know them:

> I learned a lot about making assumptions when it came to managing contractors. In one situation I had assumed that this contractor was capable of doing [a project], and it was a situation where basically that project didn't get done and had to be postponed. So, what I learned was the

more information you can give them, the more successful you'll be. (Food Manufacturer, Chemical Engineering major, eight months on the job)

I learned that the people out on the floor are kind of the eyes and ears of the operation. They observe firsthand what's happening. So you better get to know them well and listen to what they have to say. Earning their respect is also important because if they don't think that you know what's going on, then they're going to have a hard time discussing their problems with you. (Computer Parts Company, Mechanical Engineering major, seven months on the job)

Working with other people was another important professional skill that young engineers reported they learned mostly on the job. Many considered having good interpersonal skills to mean getting to know coworkers or supervisors personally, establishing relationships with them, networking to gain visibility and identify key resources, and being able to convince collaborators of their ideas.

I've learned where people's strengths are. Some people are really detail oriented, so if I have a detail-oriented question, I'll go to them. Some people are better communicators. And there are people that know their stuff but when they try and explain it to you, I'm just like, "You're not even understanding my question." So I've sort of flagged those people in my mind. (Computer Parts Company, Chemical Engineering major, eight months on the job)

[I learned] that you've really got to network and get to know people on a personal level and earn their respect and respect them. It's really different around here because no one has to do anything. If you want people to go outside of the box to help you, you've got to [get to] know them. (Car Company, Mechanical Engineering major, twelve months on the job)

On the job, young engineers also acquired tips on good group work practices such as following up emails with face-to-face or phone conversations, requesting that deliverables be sent by a certain date, and checking on the status of deliverables at regular intervals. Several mentioned learning to keep others updated on the status of their own deliverables, since it prevented conflict and made it easier to locate help when problems occurred:

I learned that you've always got to check up on people and make sure that they have it [their deliverable] done in a week or so of what they said. You want to rely that they're going to do what they say they're going to do but you really can't. So you've always got to double check behind someone to make sure that he's doing exactly what they asked. (Car Company, Mechanical Engineering major, four months on the job)

I guess it's best to keep everyone in the loop always. Send out mass emails to try to keep everyone in the loop, so then there isn't questioning. Everyone knows where you're at with it, and then if they have an idea, they can throw it out and you can try that idea too. (Computer Parts Company, Mechanical Engineering major, seven months on the job)

Young engineers also reported having to learn information-finding skills on the job. For a few, information finding meant becoming familiar with using their organization's databases, but, as Korte (2009) documented, many more discovered that their coworkers were the most significant source of information. Some interviewees spoke of specific strategies for seeking help, such as first building camaraderie with their coworkers and then making sure not to monopolize their coworkers' time. Other professional knowledge and skills young engineers used in their work included project management skills, time-management skills, leadership skills, documentation skills, context knowledge, and a work ethic. Participants learned all but project management skills primarily or exclusively in the workplace.

Two-thirds of the participants reported applying knowledge about their organization in their technical work. A majority of those who applied this knowledge reported learning this organizational knowledge on the job (56 percent), although some (16 percent) reported acquiring the knowledge as students, particularly during internship and co-op experiences. Among all who mentioned knowledge about their organization, over 40 percent equated this knowledge with routine or bureaucratic procedures, specific policies, and procedures that they had learned (for example, how to write test protocols or submit work orders). By contrast, only 16 percent talked about the organization's background or culture. Just over a quarter said they had developed an understanding of the hierarchy or division of labor, including how different departments worked together and what role the various departments played in the bigger picture.

Indicative of the company-specific nature of organizational knowledge, only a quarter of participants who had internships mentioned what they had learned about their organizations or life in their organizations during that time, and few reported transferring that knowledge to their current job. In most cases, it was not until the interviewer asked them about what they had learned about organizational knowledge from their internships that they spoke of this topic. Perhaps these young engineers were aware of their organizations and how they operated but did not consciously apply that knowledge in their work. Our asking participants about knowledge and skills they used while doing specific tasks may have also focused their responses away from what they learned in other work settings. But some engineers did report learning about how organizations work in an internship and applying that knowledge in their work, even if they had interned with a different employer than their current one. This suggests that lessons from working in one professional engineering environment may transfer to others.

In sum, engineers reported early in their careers employing a variety of technical, professional, and organizational knowledge and skills. Technical work required knowledge of specific content, tasks, and processes, as well as general problem-solving skills. Professional work involved communicating with many different groups of people, as well as interpersonal, relational, and information-finding skills. Organizational work included understanding the culture, values, and operating procedures of their organization. Although these young engineers had developed strong technical backgrounds in school, which they expanded and refined at work, they learned most of their professional knowledge and skills, as well as their organizational knowledge and skills, at work. Additionally, although they might have gained some professional and organizational knowledge and skills through their internship or co-op experiences, most could not explicitly connect this to their technical work. The findings suggest that engineering graduates might benefit from better instruction in the professional and organizational aspects of their work while in school.

4.3.2 Changing Views of Engineering Practice (the Longitudinal Sample)

Comparing the knowledge and skills that engineers thought were important when they were college juniors with those they thought were important after four years on the job provides a sense of how engineers' perceptions change as they move from being students to being practitioners. Table 4.5 shows the number of Longitudinal Sample participants who mentioned various technical and professional skills or areas of knowledge at both points in time.

Table 4.5 shows that participants' impressions of the general importance of professional knowledge and skills remained relatively constant over time, but the relative importance of technical knowledge and skills declined with twenty fewer mentions after working for four years. Respondent engineers were less likely to mention math, logic, science, problem-solving, and visualization skills as crucial for their work. They were just as likely to mention content knowledge and software skills, and like engineers in the Workplace Sample, provided specific examples of knowledge and applications once in a job. The pattern suggests that engineering students may mistakenly overemphasize the relevance of broad (versus specific) technical knowledge and skills to engineering practice.

This supposition is supported by responses that interviewees gave when asked to define engineering as students. Seven described engineering as problem solving, and three explicitly emphasized that engineering entailed the application of math and science.

Engineering is the applied use of science and technology to solve problems. (Technical Public Institution, Engineering Physics major, junior year)

Table 4.5   **Comparison of important knowledge and skills (Longitudinal Sample,**
n **= 9)**

	Number of participants who reported skill/ knowledge as important		
	Junior year of college	Four years postgraduation	Difference
Technical knowledge and skills[a]	9	6	−3
Math skills	8	1	−7
Logic skills	5	0	−5
Science skills	4	0	−4
Problem-solving skills	3	2	−1
Technical skills (general)	2	1	−1
Visualization skills	1	0	−1
Content knowledge	2	2	0
Software skills	1	1	0
Total no. mentions	*26*	*6*	*−20*
Professional knowledge and skills	8	7	−1
Context knowledge	3	0	−3
Creativity skills	3	0	−3
Teamwork skills	2	0	−2
Social skills (general)	1	0	−1
Self-directed learning skills	1	1	0
Business knowledge	0	1	+1
Documentation skills	0	1	+1
Self-motivation skills	0	1	+1
Working with people	1	3	+2
Communication skills	2	5	+3
Total no. mentions	*13*	*12*	*−1*

[a] Number of participants who reported using/learning at least one of the technical knowledge and skills listed in the table. The number of participants who reported using/learning professional knowledge and skills can be interpreted similarly.

> Engineering is the art of figuring out interesting problems that need to be figured out using math and science. (Suburban Private University, Computer Science major, junior year)

> Engineering is just coming up with solutions to problems using math and science. (Technical Public Institution, Petroleum Engineering major, junior year)

Furthermore, when asked as students where they learned the knowledge and skills they considered important, the participants said they learned math and science in the classes they had been taking since elementary school. A few mentioned teachers that had encouraged them to pursue engineering because of their abilities in science and math. Thus, the students' images of engineering revolved around technical knowledge and skills by its link to math and science in school.

My childhood was pretty exposed to gaining those types of skills. My father was an accountant so his math background helped a lot. As far as grade school, I always had teachers that were supportive with math and science in general. So I was kind of encouraged in that area. (Large Public University, Computer Engineering major, junior year)

By contrast, when students said they learned nontechnical knowledge and skills, the learning was linked to experiences outside of school, such as extracurricular activities and sports.

Being social is easy for me. I'm involved in a whole bunch of different things on campus that allows me to get out there and meet new people. That's how I developed that one [social skills]. (Large Public University, Civil and Environmental Engineering major, junior year)

The findings also suggest that even though participants ascribed some importance to professional knowledge and skills when they were students, they were not attuned to those most needed in practice. As students, the engineers thought teamwork skills, context knowledge, and creativity skills were most important. Knowledge and skills such as communication and working with people were deemed more valuable four years later, when mentions of these two skills more than doubled. The meaning of these skills also changed from very general communication and teamwork skills as juniors to interacting with people within and outside their work group as working engineers. This change is consistent with the evidence from the Workplace Sample in which young engineers spoke of needing to work with different groups of people, but not necessarily in teams. This could be another case of students aligning their beliefs of what is important to engineering practice with what they experienced in the classroom. Or, to put it differently, engineering programs may overemphasize teamwork skills and underemphasize communication skills, especially the ability to communicate with people from different positions, disciplines, and even walks of life.

I work with a lot of different groups of people, and I communicate with a lot of different types of people, and the way that I communicate with my operators is very different than the way that I communicate with the engineers that I work with, just because of their level of education and understanding of certain things. (Technical Public Institution, Chemical Engineering major, four years postgraduation)

I have to communicate recommendations and changes to a wide variety of people, from people with engineering experience in my own group to operators who have a high school education to operating management. So being able to communicate to a wide array of people is very important. (Technical Public Institution, Chemical Engineering major, four years postgraduation)

Nevertheless, despite many similarities, participants in the Workplace Sample and in the Longitudinal Sample had somewhat different opinions of the knowledge and skills that engineers require. Participants in the Longitudinal Sample did not mention specific job tasks and processes, information-finding skills, or knowledge about their organizations. Perhaps this reflects the fact that the studies framed the issue differently; the workplace study asked engineers about task-specific knowledge and skills used to work on a particular project or problem, while the longitudinal study asked about metaskills important to their work in general. Alternately, participants in the Workplace Sample spoke about knowledge and skills used in the first couple of years at work, while longitudinal participants spoke about knowledge and skills at two very different points in their career. The two groups of engineers could have differed in what knowledge and skills were important because participants in the Longitudinal Sample had been working longer than participants in the Workplace Sample. As engineers work longer, they might move into less technical roles with broader scope. Some knowledge and skills may become less important to engineers' work, causing these aspects of the work to recede into the background.

Strikingly, no engineer in the Longitudinal Sample specifically identified project management skills as important, either as students or after four years on the job. Yet, when asked whether their idea of an engineering job had changed since they graduated, six of the nine participants said that they had not realized their work would include project management including such tasks as dealing with people, attending meetings, writing documents, and creating schedules and budgets. Given the high frequency with which some interviewees performed these tasks, it is unclear why they did not identify project management skills as important. One possible reason is that they did not perceive project management to be "real engineering," a finding corroborated by the Workplace Sample. With regard to how their idea of an engineering job had changed, two longitudinal participants replied:

> There are parts of the job that I don't think make use of my engineering skills but I think are necessary for the role that I'm in, things like writing procedures. I don't think that's necessarily what I would call an engineering task, but sometimes it's necessary just because I'm the one with the knowledge required to write that procedure. (Technical Public Institution, Chemical Engineering major, four years postgraduation)

> I think a lot of it [work] can be time management and dealing with people. But as far as strict engineering like what we were taught in school—calculations and stuff like that—it's a lot of fun but it doesn't take up the majority of my day. (Technical Public Institution, Petroleum Engineering major, four years postgraduation)

Further analysis of our interviews with the Longitudinal Sample suggests that students may have unrealistic or distorted views of engineering. As students, participants saw engineering education and practice as primarily technical, and had overly narrow perceptions of the professional aspects of engineering. Some of their misconceptions appear to remain with them into the workplace, influencing their views of their jobs. To the degree that students either choose or do not choose engineering as a profession because of these views, the profession has a problem. Moreover, those who choose engineering may not be taking advantage of opportunities at school that would make them better engineers at work.

4.3.3   Changing Expectations for Engineering Practice
(the Workplace Sample, Revisited)

With the changes in how our longitudinal participants perceived engineering in mind, we return to the Workplace Sample to compare students' expectations for professional practice with the realities they faced as new hires. We asked the fifty-seven engineers in the Workplace Sample to tell us how engineering practice differed from the expectations they had when they were in school. Some claimed there were no differences and that their classes, extracurricular activities, and internships had well prepared them for work. A few considered school and work to be similar, since both required using the same tools and concepts to solve problems:

> I would say [school is similar] in that you're given a problem and left to go from there. You've got to figure out what exactly you need to do, how you're going to go about doing it, and then get the results and look at them and figure out if they are good. I feel like that's similar to the work here. (Car Company, Mechanical Engineering major, fifteen months on the job)

> It's easier to answer how they're [school and work] the same, actually. I don't use the equations for almost anything anymore, but I understand the concepts and I use those all the time. (Computer Parts Company, Mechanical Engineering major, eight months on the job)

More, however, noted differences, as we have pointed out in prior writings; for example, engineers' work emphasizes application while school emphasizes theory, problems at work have greater scale and complexity, and at work, engineers need to seek out information actively on their own (Korte, Sheppard, and Jordan 2008; Brunhaver et al. 2010). Particularly telling is that young engineers quickly learned that professional practice requires more social interaction than technical work, and that it is more cross-disciplinary than school led them to appreciate.

Of the twenty-nine participants who said that engineering work was more social than technical, sixteen admitted being surprised by this. Prior to starting their jobs, they expected primarily technical work. For some, this meant spending considerable time doing design or analysis. Others expected to use the equations and theories learned in school. But once on the job our interviewees reported spending more time than they expected managing people, helping to troubleshoot equipment, and working on the production floor. For those who had expected to work alone, the reality was difficult to accept:

> Yeah, that [doing more technical work] was my expectation and that's why I chose to be a civil engineer. I just want to do technical work and don't [want to] deal with people and management. (Transportation Department, Civil Engineering major, twelve months on the job)

> I think there's a lot of disillusion once you leave engineering school, because when you're there, you're taking all these tests, you're doing all these math problems, and you're delving deep into theories and equations and formulas, and I haven't touched any of that stuff in the year and half that I've been here. (Food Manufacturer, Chemical Engineering major, seventeen months on the job)

Some of the young engineers had theories as to why their work was so different. One believed that his or her organization had already solved all of the hard problems and now only needed engineers to verify the results.

> Engineering here seems less technical at times because the company is established and the processes have been established. There's already been a lot of testing. So we have learned everything. I'm dealing a lot more with tests now, verifying I guess what the computer programs are getting out. (Computer Parts Company, Mechanical Engineering major, nine months on the job)

Another believed that their organization outsourced technical work to save time and money.

> What I learned was that we typically, at least at this plant, don't do so much [design]. We are resource limited, so rather than spend our time trying to figure things out, you know, "I need a two-inch pipe," we farm that out to an outside vendor. (Food Manufacturer, Chemical Engineering major, seventeen months on the job)

Still others thought the amount of technical work varied widely by position, and that even though they were not doing as much technical work, some of their coworkers could be. Besides, no matter how technical the job, engineering work would always involve some social interaction.

> There's technical [work] and project management, but in the particular role I'm in now, it's more project management than it is technical. And you'll find that within [organization], it depends on the position that you're in. (Transportation Department, Civil Engineering major, thirty-six months on the job)

> Yeah, there are some places you can go technical. But there is nothing that's centrally technical. There has to be communication, you have to work with people. I think that's the major thing I learned when I started at [organization]. (Transportation Department, Civil Engineering major, twelve months on the job)

Sixteen participants said engineering work required them to reach beyond the discipline they studied in school. Four of the sixteen were not surprised. Even in school they had expected to be primarily general problem solvers and only secondarily, disciplinary experts.

> I was studying to be an electrical engineer, but [I learned] I was going to be an engineer first, and then an electrical engineer. Engineering has to do with solving problems, making requirements, stuff like that. I learned that that's how engineering was going to be, and I've come to the conclusion that, yes, that's how it is. (Car Company, Electrical Engineering major, five months on the job)

The other twelve were surprised their employer made them responsible for projects outside their expertise. At Car Company a few mechanical engineers found themselves doing electrical or computer engineering-related work. Similarly, chemical engineers at Food Manufacturer and Computer Parts Company became involved in mechanical and civil engineering-related work. Some interviewees found the differences between their work and the discipline they studied to be nuanced, but others, like the chemical engineers at Food Manufacturer and Computer Parts Company, were not working with any of the processes or equipment learned in school and felt thrown into a completely different field:

> I would have never guessed that I'd be making circuits. I never thought about how the chemical engineering process can be applied in that type of scenario. This company does not have actual chemical engineering processes. I wasn't really expecting [that] big a contrast. (Computer Parts Company, Chemical Engineering major, six months on the job)

Overall the reflections of the engineers in the Workplace Sample suggest that engineering work is much more variable, complex, and social than most engineering curricula convey. The picture is consistent with the other data we have discussed. A young engineer's work is less about using theories or equations than about project management and working with other people. Engineering practice is not confined to a single area of expertise. Instead, it requires young engineers to pick up new knowledge and skills on the job. Given the reported gap between work and education, it seems reasonable to suggest that engineering programs and engineering practice be reconfigured to achieve closer alignment between school and work. We elaborate on this idea next.

4.4   Summary and Implications

Our interviews with early career engineers in the Workplace and Longitudinal Samples point to two distinct but interrelated sides of engineering practice: the technical and professional sides. In addition to technical work, young engineers are responsible for nontechnical tasks that require significant social interaction, such as managing projects and coordinating the work of other people. Employers also expect young engineers to be able to work outside the specific discipline in which they trained and to work with people who were not engineers.

Despite these realities, engineering students emerge from their programs with relatively narrow views of professional practice. Although students may ascribe some importance to nontechnical skills, they mainly conceive of engineering as technical problem solving involving the direct application of theory and equations they learned in classes. Students develop these conceptions in part by looking at their teachers and college professors. Most respondents in a recent survey of K–12 educators associated engineering with math, science, and making and fixing things (Yasar et al. 2006) while Pawley (2009) showed that engineering faculty described their profession in much the same way. Given that students learn and internalize these messages as early as elementary school (Capobianco et al. 2011; Cunningham, Lachapelle, and Lindgren-Streicher 2005; Oware, Capobianco, and Diefes-Dux 2007), it is not surprising that they persist beyond formal schooling.

To be sure, the data indicate that the technical knowledge and skills that students learn in school are indeed important to engineering practice. Furthermore, most young engineers continue to refine and expand their knowledge and skills after starting work. This later learning, however, focuses primarily on the doing of a specific job. Once on the job, the importance of technical knowledge and skills appears to decline. Many engineers in the Workplace Sample even noted the lower importance of these competencies from the start.

But as technical knowledge and skills become less central or less sufficient for doing engineering work, the importance of professional knowledge and skills increases. Even when students are exposed to the professional side of engineering work in school, they may not fully grasp what this work looks like on the ground. This seems particularly true for communication and teamwork. Although some students think that communication skills are important to engineering, it is only after they start working that they begin to speak explicitly about the value of formal (technical reports and oral presentations) and informal (e.g., interacting with others via phone and email) communication on the job. Similarly, students readily talk about the importance of working in teams, but none of our young, employed engineers described teamwork in the manner typically found in engineering programs (Colbeck, Campbell, and Bjorklund 2000). Instead, they spoke of working and communicating with different groups of people, including other engineers, operators, managers, clients, and suppliers. This discrepancy is systemic of a larger issue in engineering education in which faculty lack the time, resources, and incentives to create multidisciplinary experiences within the current disciplinary structure (Jamieson and Lohmann 2009; McNair et al. 2011).[8]

The data show that once on the job, young engineers' appreciation of professional knowledge and skills becomes more nuanced. Yet, engineers are still apt to consider less technical work, such as project management, to differ from "real engineering" and not to be directly related to their degrees. Going forward, outsourcing and automation will likely make project management a common position for engineering graduates. Since students expect synergy between "what is learned in [the] classroom and what is needed in the field for successful practice" (Steering Committee of the National Engineering Education Research Colloquies 2006, 259), this affects how engineers value their education. Finally, two-thirds of the engineers in the Workplace Sample mentioned knowledge they had learned about their organizations in order to do their jobs. However, most participants failed to see connections between this knowledge and their technical work on their own. Few of those who had held undergraduate internship or co-op experiences could transfer lessons they had learned then to their current work situations now. These findings substantiate employers' low ratings of engineering graduates in the area of organizational contexts and constraints (see introduction; Lattuca, Terenzini, and Volkwein 2006). It is also puzzling that none of the participants in the Longitudinal Sample mentioned knowledge about their organization as important, especially since knowing how an organization operates and how to successfully negotiate hierarchies, divisions of labor, and status structures are critical for success in almost any organization. We suspect, however, that young engineers fail to mention organizational policies and processes because they do not see them as a form of knowledge unless explicitly cued to do so.

Our findings are consistent with and extend prior studies of engineering education.[9] Salzman (2007) reported that because managers find technical skills to be common, they cannot use them to distinguish between job can-

---

8. Notwithstanding the recent focus on such teamwork (ABET 2011; National Academy of Engineering 2004), most group work is done in students' own disciplines, and few faculty members have experience or training in managing groups (Colbeck, Campbell, and Bjorklund 2000).

9. Bucciarelli and Kuhn (1997) argued that formal education prepares engineers to succeed in the "object world" (211) but overlooks the process-oriented, context-laden social world. Based on a study of engineers in a large high-tech company, Perlow and Bailyn (1997) not only concurred with Bucciarelli and Kuhn's assessment, but added that engineers typically perceive "real engineering" to pertain only to the object world. Through semistructured interviews with a broad sample of engineers, Trevelyan (2009, 2010) found that most engineering curricula focus solely to the technical aspects of engineering even though engineering is both a technical and social discipline. Similarly, Sheppard et al. (2008) described engineering as interactive and complex work that encompasses many domains beyond the technical. Sheppard et al. also found such an image of engineering contrasted sharply with the narrow way that most engineering education is currently framed.

didates. Managers seek young engineers who possess nontechnical skills, especially communication skills and the ability to work across "borders," both disciplinary and organizational. Yet, these skills are often the most difficult for the managers to find in new engineers (Salzman and Lynn 2010; Lattuca, Terenzini, and Volkwein 2006).

These studies point to deficiencies in the current model of engineering education that constrains it from producing engineers with knowledge and skills required for being effective in the workforce and suggests the value of elevating professional and organizational knowledge and skills in training engineers. Complicating such a change is the fact that most engineering programs are burdened by a crowded curriculum, which makes adding content a challenge (Jamieson and Lohmann 2009; Salzman and Lynn 2010; Trevelyan 2007). Moreover, because cutting edge technical training is central in engineering education, any redesign of engineering pedagogy, assessment, and accreditation to connect education with practice will have to find ways to integrate professional and organizational knowledge and skills with training in the latest engineering concepts and tools (Froyd and Ohland 2005; Litzinger et al. 2011; Sheppard et al. 2008).

4.4.1   Implications for Engineering Education

To develop competent and productive graduates, engineering schools should consider ways to emphasize the professional and organizational content as well as technical skills—not only in the classroom but in cocurricular and extracurricular activities as well. Many engineering students hold the unrealistic view that engineering is synonymous with technical problem solving even after they have completed design projects (such as senior capstone) in upper-division courses. Given how resistant to change the image of engineering as solely technical seems to be, engineering schools should seek ways to help students connect their early experiences in math and science to later engineering experiences and, ultimately, to professional practice. Providing opportunities to learn about real engineering work at every stage of an undergraduate career or even as part of K–12 education could potentially improve the attractiveness of engineering as a career. One way to accomplish this goal would be through cognitive apprenticeships (Collins, Brown, and Holum 1991; Sheppard et al. 2008) that expose students to professional practice through carefully staged and monitored steps. In this approach, educators first model expert practice; then, they scaffold students' efforts to imitate their performance, providing feedback where needed. The process is repeated over time, moving from simple lab and design exercises in the students' freshman year to closer "approximations of practice" (Grossman et al. 2005) by their senior year.[10]

10. Other ways to expose students to real engineering include pedagogies designed to help students make sense of and develop the abilities for practice. Examples include design tasks

Because engineering graduates work with a wide variety of people including nonengineers, collaborative and cooperative learning, which emphasize small group work, are particularly promising (Johnson, Johnson, and Smith 1998; Prince 2004; Smith et al. 2005) especially when the groups are demographically and disciplinarily heterogeneous so that students learn how to work with people who are different from themselves (Colbeck, Campbell, and Bjorklund 2000; McNair et al. 2011). Similarly, students could be required to take courses in other departments to prepare for multidisciplinary products and projects in the workplace and engineering programs and schools could support faculty to make this happen (Jamieson and Lohmann 2009; McNair et al. 2011).

Engineering students also need to learn to communicate their ideas to a variety of audiences and in many modes. For this objective, project-based learning may be particularly useful (Dym et al. 2005). When implemented in teams, projects allow students to practice formal communication via technical reports and oral presentations as well as informal communication through email, memoranda, group meetings, and so on. Projects also expose students to other professional knowledge and skills including the management of tasks, schedules, and people. Finally, engineering graduates must be able to direct their own learning when they recognize they do not know something. Problem-based learning requires students to formulate their own problems and then find information for solving the problems (Prince 2004; Woods 1994). Students also need to begin becoming savvy about organizations while in school. Given the wide range of paths that students eventually pursue, no curriculum can address every detail of an organization's history, culture, policies, and procedures, but it is possible to expose students to a variety of organizational systems through field trips, case studies, and in-class speakers and to design assignments to emphasize workplace contexts and constraints that mimic those that students might encounter. Engineering students could be encouraged to take courses in organizational behavior and other related topics. Even technical-engineering courses could include a stronger emphasis on the organizational and contextual influences that affect the practice of engineering.

Engineering students might benefit from opportunities to learn about professional practice outside the classroom. Venues for doing so would include research experiences, study abroad, involvement in professional societies, internships, co-ops, and other forms of employment. Internships and co-op experiences provide firsthand insight into what engineers really do and how they use the knowledge and skills they learned in school and a head start on

---

and laboratory work (Litzinger et al. 2011; Sheppard et al. 2009). More recently, educators and researchers have experimented with other methods such as portfolios (Dunsmore, Turns, and Yellin 2011; Eliot and Turns 2011) and think-aloud protocols (Douglas et al. 2012). Given their emphasis on the social, these latter examples may be especially effective for teaching professional and organizational knowledge and skills.

developing a professional network. Students who participate in internships and co-ops are better prepared for the workplace and, thus, more employable (Haag, Guilbeau, and Goble 2006). Internships and co-ops may also help students more knowledgably choose an engineering discipline or decide whether to pursue an engineering career (Raelin et al. 2011). Engineering programs could also actively help students reflect on and integrate these experiences into their understanding of engineering practice.

Incorporating such changes into a four-year engineering curriculum will not be easy. Some academicians, organizations, and even some students have called for either extending the engineering bachelor's degree to five years or making the master's degree the first professional engineering degree to attain a more equal balance of theory and practice (National Academy of Engineering 2005; Grose 2012). Revising engineering curricula may require considering how to redesign traditional courses to more closely align with practice or by building consensus on what conceptual and theoretical knowledge is more relevant for practice.

Other stakeholders can also help tighten the connection between schooling and work (Korte 2009; Sheppard et al. 2008). Firms can partner with engineering schools to assist in redesigning programs and developing ways for students and practitioners to interact inside and outside of the classroom. Practicing engineers might deliver guest lectures, coach students, provide program feedback, and serve as adjunct faculty. Firms and trade associations can sponsor fieldtrips, design projects, internships, and co-ops and can sponsor educational innovation and scholarship both in the university and industry (Jamieson and Lohmann 2009). National research, accrediting, and professional organizations play a vital role in advocating for educational reform. They promote and reward engineering programs and faculty dedicated to experimenting with innovative pedagogies, and endorse best practices through assessment and accreditation, and thus could help engineering schools become more practice oriented (Sheppard et al. 2008).

Finally, our study and others suggests that, much like collegiate engineering programs, K–12 education tends to emphasize the importance of math and science to engineering. Others have argued that primary and secondary schools are neither teaching the right knowledge and skills nor sending the right messages to students about engineering (National Academy of Engineering and National Research Council 2009; Anderson et al. 2011; Capobianco et al. 2011). For example, K–12 students often believe that engineering is sedentary work that involves little interaction with people (National Academy of Engineering 2008a); K–12 educators who expose students to images of how engineers use their skills could also reduce stereotypes that students have about engineering as well.

4.4.2   Implications for Engineering Practice

Industry can help young engineers adjust more effectively to the transition from school to work. Particularly helpful would be information at the point of hiring, if not sooner, about the kind of work a student will be doing. During the first few months of a job, discussions of the employer's culture, history, policies, and procedures could be contextualized so that new recruits can envision how these factors will influence their work. Young engineers would benefit if managers and coworkers served as mentors to help them see how the knowledge and skills that they learned in school apply to their jobs, particularly if they will be doing project management (Korte 2009, 2010). Coaching and mentoring enhances internships and co-op experiences by giving students exposure to diverse people and organizational settings and more realistic job previews (Parsons, Caylor, and Simmons 2005; Raelin et al. 2011).

To achieve these objectives, employers need to better understand the work that they hire young engineers to do, so that they can set their own expectations appropriately. In many firms, a bachelor's degree is a requirement for practice, yet some engineering jobs can be adequately filled by workers with an associate's degree. Employers could ask whether they are properly utilizing their young engineers and whether these individuals are capable of doing more complex and creative work. If the latter is true, then the young engineers' roles and responsibilities need to be adjusted so that they can contribute more to the organization and pursue meaningful work. Employers can collaborate more closely and meaningfully with engineering educators and researchers so that engineering curricula can be improved.

Perhaps most troubling for the profession is that some students may turn away from engineering careers because they have a limited understanding of what engineering is or because how it makes use of the knowledge and skills they value (Atman et al. 2010; Lichtenstein et al. 2009). For students confident in their communication and interpersonal skills, realizing that engineering involves a significant amount of social interaction may make engineering more attractive. Students without this confidence may find engineering less attractive. When asked how her idea of an engineering job had changed since graduation, one interviewee remarked, "I'm just seeing more of the opportunities I didn't even know existed when I was in school" (Technical Public Institution, Chemical Engineering, four years postgraduation).

4.4.3 Implications for Future Research

Additional research is needed on how engineering students' concepts of engineering work affect their careers and to what extent our finding that, over the first four years of employment or practice, the importance of general technical knowledge and skills in engineers' work wanes while the importance of professional and organizational knowledge and skills rises due to the role of shifts in engineers' work responsibilities or in their perceptions of shifts. Further study is also needed into how the nature of engineering work varies depending on the type of organization or project. Larger data sets can help, for example, determine the influence of organizational size or industry sector, or of engineering discipline and institution type. Our longitudinal study may also be expanded to capture how engineers' views change over a longer period of time. Our identifying organizational knowledge as an important competency for young engineers suggests the value of further research into the role of such knowledge and how engineers attain it in career development.[11]

Finally, the gaps between engineering education and practice that we have analyzed mirror similar gaps found in nursing, medicine, business, law, accounting, and teaching (Colby et al. 2011; Cooke, Irby, and O'Brien 2010; Jones 2002; Sullivan et al. 2007). Like engineering, programs in these professions struggle with the integration of professional knowledge and skills with technical content. Too often, the gap between what university programs teach and what employers want can also be seen as a divergence between what education provides and what firms provide.[12] Both education and industry could improve the fit by working out together their expectations for graduates and what their respective roles and responsibilities will be, with a strong involvement of students. That there are gaps between professional training and work not only in engineering but in other key professions suggests the value of future research on the experiences of professional training and work more broadly, from which engineering and the other professions can gain.

11. In our Workplace Sample, many spoke of not understanding how their work fit into the organization or not knowing what they needed to know (Brunhaver et al. 2010; Korte, Sheppard, and Jordan 2008).

12. According to Cappelli (2007) the real reason is that employers no longer provide the internal training needed to develop the skills of new hires, seeking instead new hires who can "step immediately into the job and start doing the work" (Center for Education 2008, 67), and for educators policymakers to take on the responsibility for training that they have shouldered in the past. While controversial (Center for Education 2008), these comments set the stage for a discussion between the university and industry.

Appendix

Table 4A.1   Definitions of knowledge and skills mentioned by the workplace and longitudinal samples

Knowledge and skills	Sample	Definition
Business knowledge	Longitudinal	Awareness of the organization's business needs
Communication skills	Workplace, Longitudinal	Ability to express oneself effectively, in written/oral reports and when working with different groups of people
Content knowledge	Workplace, Longitudinal	Understanding of technical content, including engineering, science, and math
Context knowledge	Workplace, Longitudinal	Awareness of contextual issues affecting engineering solutions, for example, safety, finances
Creativity skills	Longitudinal	Ability to develop original ideas or solutions
Design skills	Workplace	Ability to generate and develop concepts for a system, component, or process
Documentation skills	Workplace, Longitudinal	Ability to take good notes and keep track of records
Equipment/processes knowledge	Workplace	Understanding of equipment and processes needed to do a job
Hands-on skills	Workplace	Ability to make or build an object using tools/processes
Information-finding skills	Workplace	Ability to locate resources and information
Leadership skills	Workplace	Ability to provide direction for a team or project
Logic skills	Longitudinal	Ability to draw conclusions using reasoning
Math skills	Longitudinal	Ability to apply math knowledge and methods
Modeling and analysis skills	Workplace	Ability to model, analyze, and interpret data
Organizational culture/background knowledge	Workplace	Understanding of the organization's culture and background
Organizational hierarchy knowledge	Workplace	Understanding of the organization's labor hierarchy
Organizational policies/procedures knowledge	Workplace	Understanding of the organization's policies and procedures

(continued)

Table 4A.1 (continued)

Knowledge and skills	Sample	Definition
Problem-solving skills	Workplace, Longitudinal	Ability to define and solve engineering problems
Programming skills	Workplace	Ability to write and troubleshoot software code, for example, in C or C++
Project management skills	Workplace	Ability to plan, organize, and manage project resources
Rapid iteration skills	Workplace	Ability to try out different solutions in a rapid manner
Self-directed learning skills	Longitudinal	Ability to acquire new knowledge and skills on one's own
Self-motivation skills	Longitudinal	Ability to focus one's energy and effort toward doing one's work, without influence from other people
Science skills	Longitudinal	Ability to apply science knowledge and methods
Social skills (general)	Longitudinal	General skills required to interact with others on the job
Software skills	Workplace, Longitudinal	Ability to use software applications, for example, CAD
Systems-thinking skills	Workplace	Ability to visualize the relationships among a system's parts, as opposed to just the parts themselves
Teamwork skills	Longitudinal	Ability to function on teams
Technical skills (general)	Longitudinal	General skills required to do the technical aspects of a job
Testing skills	Workplace	Ability to plan, conduct, and collect data from tests
Time management skills	Workplace	Ability to manage one's time and meet deliverables in a timely manner
Visualization skills	Longitudinal	Ability to mentally picture an object or process
Work ethic	Workplace	Willingness to keep working until the task is finished
Working with people	Workplace, Longitudinal	Ability to get along and work with other people

References

ABET. 2011. Criteria for Accrediting Engineering Programs. Accessed Sept. 1, 2012. http://www.abet.org/wp-content/uploads/2015/04/appm-2010-2011.pdf.

American Society of Civil Engineers. 2008. *Civil Engineering Body of Knowledge for the Twenty-First Century: Preparing the Civil Engineer for the Future*, 2nd ed. Reston, VA: American Society of Civil Engineers.

Anderson, Kevin, Sandra Courter, Thomas McGlamery, Traci Nathans-Kelly, and Christine Nicometo. 2009. "Understanding the Current Work and Values of Professional Engineers: Implications for Engineering Education." Proceedings of the American Society for Engineering Education Annual Conference, Austin, TX, June 14–17.

Anderson, Kevin, Sandra S. Courter, Mitchell J. Nathan, Amy C. Prevost, Christine G. Nicometo, Traci M. Nathans-Kelly, Thomas D. McGlamery, and Amy K. Atwood. 2011. "Special Session: Moving towards the Intended, Explicit, and Authentic: Addressing Misalignments in Engineering Learning within Secondary and University Education." Proceedings of the American Society for Engineering Education Annual Conference, Vancouver, BC, Canada, June 14–17.

Atman, Cynthia J., Sheri D. Sheppard, Jennifer Turns, Robin S. Adams, Lorraine N. Fleming, Reed Stevens, Ruth A. Streveler, et al. 2010. *Enabling Engineering Student Success: The Final Report for the Center for the Advancement of Engineering Education*. San Rafael, CA: Morgan & Claypool Publishers.

Bankel, Johan, Karl-Fredrik Berggren, Karen Blom, Edward F. Crawley, Ingela Wiklund, and Soren Ostlund. 2003. "The CDIO Syllabus: A Comparative Study of Expected Student Proficiency." *European Journal of Engineering Education* 28 (3): 297–317.

Brunhaver, Samantha, Russell F. Korte, Micah Lande, and Sheri D. Sheppard. 2010. "Supports and Barriers that Recent Engineering Graduates Experience in the Workplace." Proceedings of the American Society for Engineering Education Annual Conference, Louisville, KY, June 20–23.

Bucciarelli, Louis L., and Sarah Kuhn. 1997. "Engineering Education and Engineering Practice: Improving the Fit." In *Technical Work and the Technical Workforce*, edited by Stephen R. Barley and Julian E. Orr. Ithaca, NY: Cornell University Press.

Capobianco, Brenda M., Heidi A. Diefes-Dux, Irene Mena, and Jessica Weller. 2011. "What Is an Engineer? Implications of Elementary School Student Conceptions for Engineering Education." *Journal of Engineering Education* 100 (2): 304–28.

Cappelli, Peter. 2007. *Talent on Demand*. Cambridge, MA: Harvard Business School Press.

Center for Education. 2008. *Research on Future Skill Demands: A Workshop Summary*. Washington, D.C.: National Academies Press.

Chubin, Daryl E., Gary S. May, and Eleanor L. Babco. 2005. "Diversifying the Engineering Workforce." *Journal of Engineering Education* 94 (1): 73–86.

Colbeck, Carol L., Susan E. Campbell, and Stefani A. Bjorklund. 2000. "Grouping in the Dark: What College Students Learn from Group Projects." *Journal of Higher Education* 71 (1): 60–83.

Colby, Anne, Thomas Ehrlich, William M. Sullivan, and Jonathan R. Dolle. 2011. *Rethinking Undergraduate Business Education: Liberal Learning for the Profession*. San Francisco, CA: Jossey-Bass.

Collins, Allan, John Seeley Brown, and Ann Holum. 1991. "Cognitive Apprenticeship: Making Thinking Visible." *American Educator* 15 (3): 6–11, 38–39.

Cooke, Molly, David M. Irby, and Bridget O'Brien. 2010. *Educating Physicians: A Call for Reform of Medical School and Residency.* San Francisco, CA: Jossey-Bass.

Creswell, John W. 1998. *Qualitative Inquiry and Research Design: Choosing among Five Traditions.* Thousand Oaks, CA: Sage Publications.

Cunningham, Christine M., Cathy Lachapelle, and Anna Lindgren-Streicher. 2005. "Assessing Elementary Students' Conceptions of Engineering and Technology." Proceedings of the American Society for Engineering Education Annual Conference, Portland, OR, June 12–15.

Douglas, Elliot P., Mirka Koro-Ljungberg, David J. Therriault, Christine S. Lee, and Nathan McNeill. 2012. "Discourses and Social Worlds in Engineering Education: Preparing Problem-Solvers for Engineering Practice." Proceedings of the American Society for Engineering Education Annual Conference, San Antonio, TX, June 10–13.

Duderstadt, James. 2008. *Engineering for a Changing World: A Roadmap to the Future of Engineering Practice, Research, and Education.* Ann Arbor: University of Michigan.

Dunsmore, Katherine, Jennifer Turns, and Jessica M. Yellin. 2011. "Looking Toward the Real World: Student Conceptions of Engineering." *Journal of Engineering Education* 100 (2): 1–20.

Dym, Clive L., Alice M. Agogino, Ozgur Eris, Daniel D. Frey, and Larry J. Leifer. 2005. "Engineering Design, Thinking, Teaching, and Learning." *Journal of Engineering Education* 94 (1): 103–20.

Eliot, Matt, and Jennifer Turns. 2011. "Constructing Professional Portfolios: Sense-Making and Professional Identity Development for Engineering Undergraduates." *Journal of Engineering Education* 100 (4): 630–54.

Eraut, Michael. 2009. "How Professionals Learn through Work." In *Learning to be a Professional through a Higher Education* (e-book), edited by Norman Jackson. Accessed Mar. 14, 2015. http://learningtobeprofessional.pbworks.com/w/page/15914952/How%20professionals%20learn%20through%20work.

Froyd, Jeffrey E., and Matthew W. Ohland. 2005. "Integrated Engineering Curricula." *Journal of Engineering Education* 94 (1): 147–64.

Grohowski-Nicometo, Christine, Traci Nathans-Kelly, and Kevin J. B. Anderson. 2009. "Work in Progress: Educational Implications of Personal History, Undergraduate Experience, and Professional Values of Practicing Engineers." Proceedings of the Frontiers in Education Conference, San Antonio, TX, Oct. 18–21.

Grose, Thomas K. 2012. "Steeper Ascent: Should a Master's Be the Minimum for Engineers?" *ASEE Prism* 21 (9). http://www.prism-magazine.org/summer12/feature_01.cfm.

Grossman, Pamela L., Christa Compton, Danielle Igra, Matthew Ronfeldt, Emily Shahan, and Peter Williamson. 2005. "Unpacking Practice: Decompositions and Approximations." Paper presented at the Annual Meeting of the American Educational Research Association, Montreal, Quebec, Canada, Apr. 11–15.

Haag, Susan, Eric Guilbeau, and Whitney Goble. 2006. "Assessing Engineering Internship Self-Efficacy: Industry's Perception of Student Performance." *International Journal of Engineering Education* 22 (2): 257–63.

Jamieson, Leah H., and Jack R. Lohmann. 2009. "Creating a Culture for Scholarly and Systematic Innovation in Engineering Education." ASEE Phase 1 Report, American Society for Engineering Education, Washington, D.C.

Jarosz, Jeffrey P., and Ilene J. Busch-Vishniac. 2006. "A Topical Analysis of Mechanical Engineering Curricula." *Journal of Engineering Education* 95 (3): 241–48.

Johnson, D. W., R. T. Johnson, and K. A. Smith. 1998. *Active Learning: Cooperation in the College Classroom.* Edina, MN: Interaction Book Company.

Jones, Elizabeth A. 2002. "Curriculum Reform in the Professions: Preparing Students for a Changing World." ERIC Clearinghouse on Higher Education, Washington, D.C. https://www.ericdigests.org/2003-4/curriculum-reform.html.

Knight, David B. 2012. "In Search of the Engineers of 2020: An Outcome-Based Typology of Engineering Undergraduates." Proceedings of the American Society for Engineering Education Annual Conference, San Antonio, TX, June 10–13.

Korte, Russell F. 2009. "How Newcomers Learn the Social Norms of an Organization." *Human Resource Development Quarterly* 20 (3): 285–306.

———. 2010. "First Get to Know Them: A Relational View of Organizational Socialization." *Human Resources Development International* 1:27–43.

Korte, Russell, Sheri Sheppard, and William Jordan. 2008. "A Qualitative Study of the Early Work Experiences of Recent Graduates in Engineering." Proceedings of the American Society for Engineering Education Annual Conference, Pittsburgh, PA, June 22–25.

Lattuca, Lisa R., Patrick T. Terenzini, David B. Knight, and Hyun K. Ro. 2014. *2020 Vision: Progress in Preparing the Engineer of the Future*. Ann Arbor: University of Michigan, Center for the Study of Higher and Postsecondary Education.

Lattuca, Lisa R., Patrick Terenzini, and J. Fredricks Volkwein. 2006. "A Study of the Impact of EC2020, Full Report." ABET, Washington, D.C.

Lichtenstein, Gary, Heidi G. Loshbaugh, Brittany Claar, Helen L. Chen, Kristyn Jackson, and Sheri D. Sheppard. 2009. "An Engineering Major Does Not (Necessarily) an Engineer Make: Career Decision-Making among Undergraduate Engineering Majors." *Journal of Engineering Education* 98 (3): 227–34.

Litzinger, Thomas, Lisa R. Lattuca, Roger Hadgraft, and Wendy Newstetter. 2011. "Engineering Education and the Development of Expertise." *Journal of Engineering Education* 100 (1): 123–50.

Lowell, B. Lindsay, and Hal Salzman. 2007. *Into the Eye of the Storm: Assessing the Evidence on Science and Engineering Education, Quality, and Workforce Demand*. Washington, D.C.: Urban Institute.

Matusovich, Holly M., Ruth Streveler, Ronald L. Miller, and Barbara A. Olds. 2009. "I'm Graduating This Year! So What IS an Engineer Anyway?" Proceedings of the American Society for Engineering Education Annual Conference, Austin, TX, June 14–17.

McMasters, J. H., and L. A. Matsch. 1996. "Desired Attributes of an Engineering Graduate." Proceedings of the AIAA Advanced Measurement and Ground Testing Technology Conference, New Orleans, LA, June 17–20.

McNair, Lisa D., Chad Newswander, Daniel Boden, and Maura Borrego. 2011. "Student and Faculty Interdisciplinary Identities in Self-Managed Teams" *Journal of Engineering Education* 100 (2): 374–96.

Meier, Ronald L., Michael R. Williams, and Michael A. Humphreys. 2000. "Refocusing Our Efforts: Assessing Non-Technical Competency Gaps." *Journal of Engineering Education* 89 (3): 377–94.

Miles, Matthew B., and A. Michael Huberman. 1994. *Qualitative Data Analysis*. Thousand Oaks, CA: Sage Publications.

National Academy of Engineering. 2004. *The Engineer of 2020: Visions of Engineering in the New Century*. Washington, D.C.: National Academies Press.

———. 2005. *Educating the Engineer of 2020: Adapting Engineering Education to the New Century*. Washington, D.C.: National Academies Press.

———. 2008a. *Changing the Conversation: Messages for Improving Public Understanding of Engineering*. Washington, D.C.: National Academies Press.

———. 2008b. "Grand Challenges in Engineering." Accessed Sept. 1, 2012. http://www.engineeringchallenges.org.

National Academy of Engineering and National Research Council. 2009. *Engineering in K–12 Education: Understanding the Status and Improving the Prospects.* Washington, D.C.: National Academies Press.

National Research Council. 2007. *Rising above the Gathering Storm: Energizing and Employing America for a Brighter Economic Future.* Washington, D.C.: National Academies Press.

National Science Foundation, National Center for Science and Engineering Statistics, Scientists and Engineers Statistical Data System (SESTAT). 2010. SESTAT Data Tool. Accessed Mar. 14, 2015. http://www.nsf.gov/statistics/sestat.

Oware, Euridice, Brenda Capobianco, and Heidi Diefes-Dux. 2007. "Gifted Students' Perceptions of Engineers? A Study of Students in a Summer Outreach Program." Proceedings of the American Society for Engineering Education Annual Conference, Honolulu, HI, June 24–27.

Parsons, Charles K., Evan Caylor, and Harold S. Simmons. 2005. "Cooperative Education Work Assignments: The Role of Organizational and Individual Factors in Enhancing ABET Competencies and Co-Op Workplace Well-Being." *Journal of Engineering Education* 94 (3): 309–19.

Passow, Honor J. 2012. "Which ABET Competencies Do Engineering Graduates Find Most Important in Their Work?" *Journal of Engineering Education* 101 (1): 95–118.

Patton, Michael Q. 2002. *Qualitative Research and Evaluation Methods.* Thousand Oaks, CA: Sage.

Pawley, Alice L. 2009. "Universalized Narratives: Patterns in How Engineering Faculty Members Define 'Engineering.'" *Journal of Engineering Education* 98 (4): 309–19.

Perlow, Leslie, and Lotte Bailyn. 1997. "The Senseless Submergence of Difference: Engineers, Their Work, and Their Careers." In *Technical Work and the Technical Workforce*, edited by Stephen R. Barley and Julian E. Orr. Ithaca, NY: Cornell University Press.

Prince, Michael. 2004. "Does Active Learning Work? A Review of the Research." *Journal of Engineering Education* 93 (3): 223–31.

Raelin, Joseph A., Margaret B. Bailey, Jerry Hamann, Leslie K. Pendleton, Jonathan Raelin, Rachelle Reisberg, and David Whitman. 2011. "The Effect of Cooperative Education on Change in Self-Efficacy among Undergraduate Students: Introducing Work Self-Efficacy." *Journal of Cooperative Education and Internships* 45 (2): 17–35.

Salzman, Hal. 2007. *Globalization of R&D and Innovation: Implications for U.S. STEM Workforce and Policy.* Testimony before the U.S. House Subcommittee on Technology and Innovation. Washington, D.C.: Urban Institute.

Salzman, Hal, and Leonard Lynn. 2010. "Engineering and Engineering Skills: What's Really Needed for Global Competitiveness?" Paper presented at the Association for Public Policy Analysis and Management Annual Meeting, Boston, MA, Nov. 4.

Sandelowski, Margarete, Corrine I. Voils, and George Knafl. 2009. "On Quantitizing." *Journal of Mixed Methods Research* 3 (3): 208–22.

Seering W. 2009. "A Curriculum that Meets Customers' Needs." Paper presented at the NSF workshop "Implementing the Recommendations of 5XME," Orlando, FL, Nov. 12–13.

Sheppard, Sheri D., Cindy Atman, Lorraine Fleming, Ronald Miller, Karl Smith, Reed Stevens, Ruth Streveler, Mia Clark, Tina Loucks-Jaret, and Dennis Lund. 2009. *An Overview of the Academic Pathways Study: Research Processes and Procedures.* TR-09-03. Seattle, WA: Center for the Advancement for Engineering Education.

Sheppard, Sheri D., Shannon Gilmartin, Helen L. Chen, Krista Donaldson, Gary Lichtenstein, Özgür Eris, Micah Lande, and George Toye. 2010. *Exploring the Engineering Student Experience: Findings from the Academic Pathways of People Learning Engineering Survey (APPLES)*. TR-10-01. Seattle, WA: Center for the Advancement for Engineering Education.

Sheppard, Sheri D., Kelly Macatangay, Anne Colby, and William M. Sullivan. 2008. *Educating Engineers: Designing for the Future of the Field*. San Francisco, CA: Jossey-Bass.

Sheppard, Sheri D., Holly M. Matusovich, Cindy Atman, Ruth A. Streveler, and Ronald L. Miller. 2011. "Work in Progress: Engineering Pathways Study: The College-Career Transition." Proceedings of the Frontiers in Education Conference, Rapid City, SD, Oct. 12–15.

Shuman, Larry J., Mary Besterfield-Sacre, and Jack McGourty. 2005. "The ABET 'Professional Skills'—Can They Be Taught? Can They Be Assessed?" *Journal of Engineering Education* 94 (1): 41–55.

Smith, Karl A., Sheri D. Sheppard, David W. Johnson, and Roger T. Johnson. 2005. "Pedagogies of Engagement: Classroom-Based Practices." *Journal of Engineering Education* 94 (1): 87–101.

Stake, Robert E. 2006. *Multiple Case Study Analysis*. New York: Guilford Press.

Steering Committee of the National Engineering Education Research Colloquies. 2006. "The Research Agenda for the New Discipline of Engineering Education." *Journal of Engineering Education* 95 (4): 259–61.

Sullivan, William M., Anne Colby, Judith Welch Wegner, Lloyd Bond, and Lee S. Shulman. 2007. *Educating Lawyers: Preparation for the Profession of Law*. San Francisco, CA: Jossey-Bass.

Sullivan, William M., and Matthew S. Rosin. 2008. *A New Agenda for Higher Education: Shaping a Life of the Mind for Practice*. San Francisco, CA: Jossey-Bass.

Trevelyan, James. 2007. "Technical Coordination in Engineering Practice." *Journal of Engineering Education* 96 (3): 191–204.

———. 2008. "Longitudinal Study of UWA Engineering Graduates (Class of 2006)." Accessed Sept. 1, 2012. http://school.mech.uwa.edu.au/~jamest/eng-work/long/info-htm.

———. 2009. "Steps toward a Better Model of Engineering Practice." Proceedings of the Research in Engineering Education Symposium, Palm Cove, Queensland, Australia, July 20–23.

———. 2010. "Mind the Gaps: Engineering Education and Policy." Proceedings of the Australian Association for Engineering Education Annual Conference, Sydney, Australia, Dec. 5–8.

Trevelyan, James, and Sabbia Tilli. 2008. "Longitudinal Study of Australian Engineering Graduates: Perceptions of Working Time." Proceedings of the American Society for Engineering Education Annual Conference, Pittsburgh, PA, June 22–25.

Vest, Charles M. 2008. "Context and Challenge for Twenty-First Century Engineering Education." *Journal of Engineering Education* 97 (3): 235–36.

Woods, Donald R. 1994. *Problem-Based Learning: How to Gain the Most from PBL*. W. L. Griffen: Waterdown, Ontario, Canada.

Yasar, Senay, Dale Baker, Sharon Robinson-Kurpius, Stephen Krause, and Chell Roberts. 2006. "Development of a Survey to Assess K–12 Teachers' Perceptions of Engineers and Familiarity with Teaching Design, Engineering, and Technology." *Journal of Engineering Education* 95 (3): 205–16.

Yin, Robert K. 2003. *Case Study Research: Design and Methods*, 3rd ed. Thousand Oaks, CA: Sage Publications.

II

Engineering and Innovation

5

The Effects of Scientists and Engineers on Productivity and Earnings at the Establishment Where They Work

Erling Barth, James C. Davis, Richard B. Freeman, and Andrew J. Wang

Studies of how scientific and engineering knowledge affects the economy focus on the impact of research and development (R&D) spending, and/or new ideas embodied in patents, on the productivity of firms. The majority of scientists and engineers in industry, however, do not perform research in corporate laboratories nor obtain patents that lead to commercially successful products or processes. Most scientists and engineers work in establishments that produce goods and services, on activities that are not classified as formal R&D. Although the pathway that links scientific and technological knowledge to lowering production costs or introducing new or improved products is critical to economic growth, we know little about the contribution of production-establishment-based scientists and engineers to productivity. Helper and Kuan's (2016) interviews and surveys of firms in the automobile supply chain show that engineers outside of formal R&D find ways to lower costs and develop new products/processes, often working with customers and/or production workers.

To see whether the employment of scientists and engineers at produc-

Erling Barth is a research professor at the Institute for Social Research in Norway and a research economist at the National Bureau of Economic Research. James C. Davis is an economist with the U.S. Census Bureau. Richard B. Freeman holds the Herbert Ascherman Chair in Economics at Harvard University and is a research associate at the National Bureau of Economic Research. Andrew J. Wang is a senior research associate in the Labor and Worklife Program at Harvard Law School and a research economist at the National Bureau of Economic Research.

This research was supported by National Science Foundation grant no. 0915670. Any opinions and conclusions expressed herein are those of the authors and do not necessarily represent the views of the U.S. Census Bureau. All results have been reviewed to ensure that no confidential information is disclosed. For acknowledgments, sources of research support, and disclosure of the authors' material financial relationships, if any, please see http://www.nber.org/chapters/c12689.ack.

tion establishments contributes to productivity broadly, we create a new establishment-firm-employee data set for manufacturing. We combine data from the quinquennial Census of Manufactures (CM) and the Annual Survey of Manufactures (ASM) on establishments' gross output and labor, capital, and intermediate inputs; the Decennial Census and Current Population Survey (CPS) on individual workers' occupation and education;[1] the National Science Foundation's Survey of Industrial Research and Development (SIRD) on firms' R&D employment; and the Longitudinal Employer Household Dynamics (LEHD) database that links every worker to their employing establishment. Appendix A provides details on how we link workers and establishments across data sets to construct a measure of the scientists and engineers proportion (SEP) of employment at the establishment level, and how we construct a firm-level measure of R&D employment.

We focus on manufacturing for three reasons. First, manufacturing is a lead sector in productivity growth. Between 1990 and 2016 the average annual rate of labor productivity increased at 3.5 percent per year in manufacturing compared to 2.0 percent in the entire economy.[2] Second, industrial R&D and employment of scientists and engineers is disproportionately concentrated in manufacturing. While manufacturing establishments employ 10 percent of the workforce in industry, they employ 20 percent of scientists and engineers in industry,[3] and manufacturing firms employ over 60 percent of R&D scientists and engineers in industry.[4] Third, data from the Census of Manufactures and Annual Survey of Manufactures allow us to analyze production and total factor productivity in manufacturing establishments.

A key statistical issue for our analysis is that we are able to link 17 percent of workers in our manufacturing establishments to Decennial Census or CPS data in order to identify their occupation. We estimate the SEP of employment from the sample of Decennial/CPS matched workers at each establishment. The absence of data on occupation for most of the workforce at each establishment creates measurement error in estimating SEP at the establishment. Measurement error is particularly severe for smaller establishments, and can substantially bias downward estimates of the effect of SEP on establishment outcomes in production-function analysis that take each establishment as an observation.

We address this issue in three ways: (a) restricting our analytical sample to

1. The Census of Manufactures and the Annual Survey of Manufactures distinguish between production (blue-collar) and nonproduction (white-collar) workers but have no information on the actual occupation of workers.
2. Bureau of Labor Statistics, *Labor Productivity and Costs*. https://www.bls.gov/lpc/tables.htm.
3. Bureau of Labor Statistics, *Occupation Employment Statistics*, OES Data, May 2013. https://www.bls.gov/oes/tables.htm.
4. National Science Foundation (2016), Detailed Statistical Tables, NSF 16-313. Table 57 reports 631,000 R&D scientists and engineers in manufacturing and 1,014,000 in all industries in 2013.

establishments that have at least ten workers that match to Decennial/CPS, and at least fifty in total employment, thus removing establishment observations with the largest likely measurement error; (b) estimating regressions that weight establishment observations by the number of Decennial/CPS matched workers; and (c) adjusting estimated establishment SEPs toward the overall mean SEP via a James-Stein type of adjustment that depends on the variance of the establishment SEP estimate (James and Stein 1961).

Our main finding is that there is a substantial positive relation between the SEP of workers at an establishment, and establishment productivity and employee earnings. And the estimated effect is substantially larger and better identified with corrections for measurement error.

The chapter is divided into four parts. Section 5.1 documents the phenomenon that motivates our analysis—the fact that most scientists and engineers work in goods- and services-producing establishments and engage in non-R&D work activities. Section 5.2 provides cross section and fixed effects estimates of the production-function relationship between establishment output and the SEP of employment at the establishment. Section 5.3 uses person-job-level data to provide cross section and fixed effect estimates of the relationship between the earnings of individual workers and the SEP of employment at their establishment. Section 5.4 concludes.

5.1   Scientists and Engineers at Goods- and Services-Producing Establishments

The impetus for this study is the fact that most industrial scientists and engineers work at goods- and services-producing establishments, and work in activities other than formal R&D. We document this fact with data on the number of scientists and engineers, from the person-level Current Population Survey (CPS) and American Community Survey (ACS) and from the establishment-level Occupational Employment Survey (OES), combined with data on the number of R&D scientists and engineers from the firm-level Business Research Development and Innovation Survey (BRDIS).[5] We also use data on work activities of scientists and engineers from the person-level Scientists and Engineers Statistical Data System (SESTAT) produced by the National Science Foundation.

Table 5.1 provides our estimates of the number and proportion of scientists and engineers in total, and working in R&D and non-R&D activities in 2013. Line 1 shows the total number of scientists and engineers employed in industry, based on data from the person-level CPS and ACS and from the establishment-level OES. The numbers are fairly similar. The CPS shows

---

5. The National Science Foundation sponsored Business Research Development and Innovation Survey (BRDIS) is the successor to the Survey of Industrial Research and Development (SIRD), which provides data on R&D employment for the 1992–2007 period covered in our production-function regression analysis.

Table 5.1   Number of scientists and engineers (in thousands) in R&D and non-R&D activities, all industry 2013

	CPS 2013 (person level)	ACS 2013 (person level)	OES 2013 (establishment level)
(1) Total scientists & engineers	5,319	4,886	4,751
(2) R&D scientists & engineers, BRDIS 2013	1,013	1,013	1,013
(3) Non-R&D scientists & engineers, (1)–(2)	4,306	3,873	3,738
(4) Non-R&D proportion of total scientists & engineers, (3)/(1) (%)	81.0	79.3	78.7

Notes: Total scientists and engineers are tabulated from CPS and ACS microdata (Ruggles et al. 2015; Flood et al. 2015) and from OES industry-occupation data (Bureau of Labor Statistics, Occupational Employment Statistics). To make the CPS, ACS, and OES figures comparable to the BRDIS figure, we include science and engineering managers in our tabulation of total scientists and engineers. Managers are 12 percent of the tabulated total in the CPS and ACS, and 11 percent of the tabulated total in the OES. Scientists and engineers are defined using Bureau of Labor Statistics, Standard Occupational Classification, Options for defining STEM occupations under the 2010 SOC, August 2012. For table 5.1, we define scientists and engineers as research, development, and design occupations and managerial occupations in life and physical science, engineering, mathematics, and information technology. R&D scientists and engineers are from National Science Foundation (2016), Business Research and Development and Innovation: 2013, Detailed Statistical Tables, NSF 16-313, tables 53 and 57. As shown in table 57, this figure is for R&D scientists and engineers and their managers. See National Science Board (2016, chapter 3), for discussion of different definitions of the science and engineering workforce and comparisons of the number of scientists and engineers. Our tabulated numbers in table 5.1 are comparable, but smaller, because we exclude social scientists and postsecondary teachers, and we cover only industry (NAICS 21-81) and exclude agriculture (NAICS 11) and government (NAICS 92).

the highest number of scientists and engineers, the ACS shows 8 percent fewer scientists and engineers, and the OES shows 3 percent fewer than the ACS.[6] Line 2 shows the number of R&D scientists and engineers from the firm-level BRDIS. Line 3 computes the number of scientists and engineers working in non-R&D activities by subtracting the Line 2 number from the Line 1 numbers. Line 4 computes the ratio of the number of scientists and engineers in non-R&D activities to the total number of scientists and engineers—about 80 percent of industrial scientists and engineers work outside of formal R&D activities.

We complement these estimates with tabulated data from the Scientists and Engineers Statistical Data System (SESTAT) on the work activity of industrial scientists and engineers. The SESTAT reports 3.808 million scientists and engineers (excluding social scientists) in industry in 2013[7]—a figure short of the figures in table 5.1. The primary reason for the lower figure is that SESTAT data exclude persons with less than a bachelor's degree. The

6. The CPS sample includes about 60,000 households per month. The ACS sample includes about 3.5 million households in each year since 2012. The OES sample for each year includes about 1.2 million establishments from a three-year period.

7. National Science Foundation (2015), *Characteristics of Scientists and Engineers in the United States: 2013*. Table 9-1 reports 4,009,000 scientists and engineers in business/industry, of which 201,000 are social scientists.

SESTAT provides information on the primary and secondary work activity of scientists and engineers, differentiating among five activities: research and development, teaching, management and administration, computer applications, and other. In 2013, 29 percent of scientists and engineers indicate that their primary work activity is R&D, and 61 percent indicate that their primary work activity is non-R&D. Tabulating both primary and secondary work activities, we find 15.3 percent of scientists and engineers indicate that both their primary and secondary work activities are R&D, 15.7 percent indicate R&D as primary activity and something else as secondary, 24.5 percent indicate R&D as secondary activity and something else as primary, and 44.5 percent indicate that both their primary and secondary activities are non-R&D.

To compute a single statistic for scientist and engineer full-time equivalent (FTE) work time engaged in R&D activities, we assume that three-quarters of a worker's FTE time is engaged in the primary work activity, and one-quarter of FTE time is engaged in the secondary work activity. With this assumption on worker FTE time allocation to primary and secondary work activity, we find that the average of scientist and engineer FTE time engaged in R&D activities is 33.2 percent, so two-thirds of industrial scientists and engineers FTE time is engaged in non-R&D activities. Since SESTAT data exclude persons with less than a bachelor's degree working as scientists and engineers, who are more likely to work in non-R&D activities compared to bachelor's degree holders, the proportion of FTE time for *all* scientists and engineers in non-R&D activities certainly exceeds the estimate of two-thirds.

Because the industry classification of workers is not comparable across the person-level CPS and ACS, establishment-level OES, and firm-level BRDIS surveys, we cannot combine data from the different surveys to estimate the proportion of scientists and engineers engaged in R&D versus non-R&D activities in disaggregated manufacturing or nonmanufacturing industries. The person-level CPS and ACS ask the respondent to classify the industry of their employer at the *location* where they work, but a worker in a large manufacturing firm may likely classify their employer as manufacturing even if they work at a nonmanufacturing establishment, such as sales or R&D or other services. The establishment-level OES classifies the industry of the establishment, and classifies workers as nonmanufacturing if they work at a nonmanufacturing establishment. The firm-level BRDIS classifies the industry of the *firm* based on the business segment where the firm conducts the most R&D, or the industry sector where the firm has the most payroll. So the BRDIS classifies scientists and engineers in R&D establishments of manufacturing firms as manufacturing workers, whereas the establishment-level OES classifies such scientists and engineers as non-manufacturing workers in the Scientific Research and Development Services industry (NAICS 5417).

To compare estimates across surveys, we tabulate the number of scientists and engineers in manufacturing from the CPS, ACS, and OES. The two person-level surveys give comparable estimates: in 2013, from the CPS we find 1.48 million scientists and engineers in manufacturing, and from the ACS we find 1.39 million. From the establishment-level OES, by contrast, we find only 0.95 million scientists and engineers in manufacturing establishments—just 64 percent of the CPS figure and 69 percent of the ACS figure.[8]

5.1.1   Scientists and Engineers Proportion of Employment at Establishments

To study the relationship between scientists and engineers, and output and productivity, at establishments, we need to estimate the SEP of employment at the establishment level.[9] To estimate SEP at establishments, we link workers in the LEHD to the 1990 or 2000 Decennial Census, or the CPS in 1986–1997, to identify the occupation of workers for the matched sample of workers. Our matched sample of manufacturing workers constitutes 17 percent of all workers in establishments observed in the Census of Manufactures or Annual Survey of Manufactures over the years 1992–2007. We use the matched sample of workers to estimate the SEP of employment at each establishment. See appendix A for further description of the data-construction procedure.

Our estimate of SEP at the establishment is subject to two forms of measurement error. The first form of measurement error relates to our measure of the occupation of workers. We identify a worker's occupation as the occupation indicated in the year that we observe the worker in the Decennial Census or CPS. If we observe a worker in more than one year of the Decennial Census or CPS, we use the observation from the most recent year, and in fact, most of our matches are from the 2000 Decennial Census. Our establishment production-function analysis covers the years 1992–2007. To the extent that workers change occupations from a scientist/engineer occupation to some other occupation, or from some other occupation to a scientist/engineer occupation, during the time period of our data, we may mismeasure the occupation of workers in our sample.

The second form of measurement error in our estimate of SEP is sampling error associated with the number of matched workers that we have at an establishment. The fewer the matches, the greater is the sampling error. The sampling error will be large in establishments with few employees, but can also be substantial in establishments with a greater number of employees.

---

8. Our tabulations for *all* workers in 2013 shows 14.2 million manufacturing workers in the CPS, 14.7 million manufacturing workers in the ACS, and 12.0 million manufacturing workers in the OES (84 percent of the CPS estimate and 82 percent of the ACS estimate).

9. The best source of occupational data for establishments is the OES, but OES establishment-level microdata are not available to us to link to census establishment-level production data. Fairman et al. (2008) show that such a link is possible.

For example, an establishment with twenty-five employees and a worker match rate of 17 percent would have information on occupation for four matched workers. If the establishment has one scientist/engineer, the true SEP is 4 percent. But with a matched sample of four workers, the estimated SEP would be 0 percent or 25 percent. For our establishment production-function regression analysis, we take three different approaches to address sampling error in our estimate of SEP at the establishment. First, we focus our regression analysis on a restricted sample of establishments with at least ten matched workers[10] and with total employment of at least fifty employees, thereby purging the sample of observations with potentially huge measurement error. Second, we estimate weighted regressions with observations weighted by the square root of the number of matched workers at the establishment. Third, we apply a James-Stein-type adjustment to the estimated SEP that shrinks the estimated SEP for an establishment toward the mean value of SEP over all establishments, with the shrinkage factor depending on the variance of the estimated SEP at the establishment.

While we focus on SEP as our key independent variable, we also use the matched-worker sample to estimate the average years of education of workers at the establishment level. This allows us to differentiate between the SEP of workers at the establishment, and average years of education of workers at the establishment, in our production-function regression analysis. The measurement-error issues relating to estimating the SEP also apply to the estimation of the average years of education of workers at the establishment. Our approaches to address sampling error in estimating SEP apply similarly to our estimate of the average years of education of workers at the establishment.

5.1.2   Manufacturing Establishments Data Set

Table 5.2 shows the mean value of selected variables for the full sample of all manufacturing establishments in our CM/ASM data, for the "matched" sample of establishments with one or more workers matched to the Decennial/CPS, and for the "restricted" sample of establishments with ten or more workers matched to the Decennial/CPS and with total employment of fifty or more. Appendix A describes the construction of our analytical data set for manufacturing establishments.

The full sample includes over 1.3 million establishment-year observations over the period 1992–2007. The matched sample includes 506,800 establishment-year observations, and the restricted sample has 215,800 establishment-year observations. The mean values of log gross output and log employment in the matched sample are somewhat larger than in the full sample, while these mean values in the restricted sample are substantially

---

10. The count of matched workers from the LEHD is for a pseudoestablishment defined by EIN-state-county-industry, so it may not be exactly comparable to total employment of the establishment from the Census of Manufactures or Annual Survey of Manufactures.

Table 5.2   Mean value of selected variables, manufacturing establishments (1992–2007)

	Establishment sample			Restricted sample of establishments, by R&D status of firm	
	Full	Matched	Restricted	R&D	Non-R&D
Number of establishment-year observations	1,305,600	506,800	215,800	128,300	87,500
Ln(gross output)	8.70	8.90	10.40	10.82	9.78
Ln(employment)	3.74	3.93	5.18	5.42	4.83
Production worker share of employment	0.724	0.723	0.729	0.716	0.747
Establishment in R&D firm	0.343	0.341	0.594	1	0
R&D scientists and engineers proportion of employment at the firm	0.016	0.016	0.027	0.046	0
Scientists and engineers proportion of employment		0.029	0.038	0.050	0.021
Scientists and engineers and science/engineering technicians proportion of employment	0.048	0.048	0.061	0.078	0.036
Average years of education of workers	11.8	11.8	11.9	12.2	11.5

Notes: The "full" establishment sample is all manufacturing establishments observed in the Census of Manufactures or Annual Survey of Manufacturers during 1992 to 2007, with total employment of five or more. The "matched" establishment sample is all manufacturing establishments with one or more workers from the LEHD that match to the Decennial Census (1990, 2000) or Current Population Survey (1986–1997), and with total employment of five or more. The "restricted" establishment sample is manufacturing establishments with ten or more workers from the LEHD that match to the Decennial Census or Current Population Survey, and with total employment of fifty or more. See appendix A for additional description. Establishment output, employment, and production worker share of employment are from the Census of Manufactures and Annual Survey of Manufacturers. R&D status of the firm, and R&D scientists and engineers proportion of employment at the firm are from the NSF Survey of Industrial Research & Development. Worker occupation and education are from the Decennial Census or Current Population Survey. Scientists and engineers, and science/engineering technicians, are defined using Bureau of Labor Statistics, Standard Occupational Classification, Options for defining STEM occupations under the 2010 SOC, August 2012. We define scientists and engineers as research, development, and design occupations in life and physical science, engineering, mathematics, and information technology. We define science/engineering technicians as technologist and technician occupations in life and physical science, engineering, mathematics, and information technology.

larger. The proportion of establishments that belong to R&D-performing firms, and the mean value of R&D worker share of firm employment, are similar for the matched sample and full sample and larger in the restricted sample. The average production-worker share of employment at the establishment level is similar in all three samples.

Using the matched-worker sample, we produce our measure of the SEP of employment at the establishment. The mean value of SEP is 0.038 in the restricted sample, which is our main regression sample. For comparison, we also produce a broader measure of the scientists and engineers and science/engineering technicians proportion of employment at the establishment. The mean value of this measure in the restricted sample is 0.061. In our regression analysis, we find that estimates using the broader measure of scientists and engineers and science/engineering technicians are similar to our main estimates using the SEP variable. Table 5.2 also presents the mean of average years of education of workers at establishments.

The last two columns of table 5.2 compare establishments in R&D and non-R&D performing firms. Previous studies find that R&D (usually measured as a stock of knowledge by accumulating R&D spending over time) is associated with higher productivity (Griliches 1998; Hall 2005; Hall, Mairesse, and Mohnen 2009). In our data, establishments in R&D-performing firms have higher gross output, higher employment, lower production worker share of employment, higher average years of education of workers, and higher SEP of employment.

5.2   Scientists and Engineers in the Establishment Production Function

If scientists and engineers at production establishments help implement technical advances that increase productivity through improved production processes or improved products, then in our manufacturing establishments data set we expect to find that the SEP of employment at the establishment is positively associated with establishment productivity. We estimate the following establishment production-function regression model:

(1) $Ln\,\text{OUTPUT} = a + b\,\text{SEP} + c\,\text{FRD} + d\,\text{FRDP} + \text{SFI}\,\gamma + \text{EMPL}\,\delta$
$+ \text{YR} + \text{IND} + \text{GEO} + u,$

where

OUTPUT is annual gross output of the establishment;
SEP is the scientists and engineers proportion of employment at the establishment;
FRD is an indicator for whether the firm is an R&D performing firm;
FRDP is the R&D scientists and engineers proportion of employment at the firm;
SFI is a vector of "standard factor inputs," including employment, capital

stock in equipment, capital stock in structures, materials, and energy (all measured in log units);

EMPL is a vector of other employer attributes, including an indicator for whether the firm is a multiestablishment firm, establishment age, establishment age squared, production worker share of employment at the establishment, and average years of education of workers at the establishment;

YR is a vector of year fixed effects;

IND is a vector of industry fixed effects

GEO is a vector of geographic region fixed effects; and

u is the error term.

With our panel data set, we also estimate an establishment fixed effects version of equation (1), which uses time variation in SEP and other variables within establishments to estimate their effect on establishment output.

Table 5.3 presents the estimated coefficients for the establishment production-function regression model. Column (1) shows the estimate using the matched sample of all establishments, with no treatment for measurement error in the variables for SEP of employment and average years of education of workers at the establishment. The positive and significant 0.079 estimate for the effect of SEP on log gross output indicates that even in the likely presence of large measurement error in SEP, establishments with higher SEP of employment have higher total factor productivity. The estimated coefficients on the standard factor inputs—employment, capital equipment and structures, materials, and energy—are all reasonable in magnitude and consistent with constant returns to scale in the production function. The estimated effect of average years of education of workers is positive, consistent with evidence that human capital is related to productivity.

The next three columns in the table show estimates using the restricted sample of establishments, that is, establishments with ten or more workers matched to the Decennial/CPS and with total employment of fifty or more. This addresses the issue of measurement error in the SEP of employment, and the average years of education of workers, by dropping observations where measurement error is likely to be exceptionally severe. This reduces the sample size of establishments by almost 60 percent, but the dropped establishments account for only 10 percent of the workers because the distribution of establishments by employment follows a power law with many establishments employing just a few workers and a smaller number of establishments employing many workers.[11] Column (2) shows that using the restricted sample more than doubles the estimated coefficient on SEP, and more than triples the estimated coefficient on average years of education of workers, compared to column (1), which indicates that measurement error is indeed a substantive issue for analysis.

---

11. See table 5.6, and the number of observations in columns (1) and (2).

Table 5.3 **Establishment production function, manufacturing establishments (1992–2007)**

Establishment sample	Matched	Restricted	Restricted	Restricted
Regression model	OLS (1)	OLS (2)	OLS (3)	Fixed effects (4)
Scientists and engineers proportion of employment	0.079*** (0.007)	0.180*** (0.019)	0.132*** (0.019)	0.055** (0.026)
R&D firm			0.063*** (0.002)	
R&D scientists and engineers proportion of employment at the firm			0.156*** (0.011)	0.048*** (0.011)
Standard factor inputs				
Ln(employment)	0.358*** (0.001)	0.354*** (0.002)	0.352*** (0.002)	0.393*** (0.003)
Ln(capital equipment)	0.061*** (0.001)	0.059*** (0.001)	0.059*** (0.001)	0.042*** (0.002)
Ln(capital structures)	0.017*** (0.000)	0.020*** (0.001)	0.019*** (0.001)	0.009*** (0.001)
Ln(materials)	0.449*** (0.001)	0.471*** (0.001)	0.470*** (0.001)	0.340*** (0.001)
Ln(energy)	0.114*** (0.001)	0.107*** (0.001)	0.105*** (0.001)	0.119*** (0.002)
Employer attributes				
Multiestablishment firm	0.094*** (0.001)	0.058*** (0.002)	0.033*** (0.002)	
Establishment age	0.005*** (0.000)	0.005*** (0.000)	0.005*** (0.000)	
Establishment age squared	−0.000*** (0.000)	−0.000*** (0.000)	−0.000*** (0.000)	−0.000*** (0.000)
Production-worker share of employment	−0.067*** (0.003)	−0.014** (0.006)	−0.014*** (0.006)	0.096*** (0.007)
Average years of education of workers	0.013*** (0.000)	0.042*** (0.001)	0.038*** (0.001)	0.004*** (0.001)
Number of observations	506,800	215,800	215,800	215,800
Adjusted R^2	0.955	0.914	0.915	0.958

Notes: Dependent variable is Ln(gross output). See table 5.2 for the definition of the "matched" and "restricted" establishment samples. The OLS models include fixed effects for year (1992–2007), industry (six-digit NAICS), and geographic region (metropolitan or micropolitan core-based statistical area [CBSA] as defined in 2009 by the U.S. Office of Management and Budget, or economic area as defined in 2004 by the U.S. Bureau of Economic Analysis). The fixed effects model includes fixed effects for establishment and year. Standard errors are shown in parentheses.

***Significant at the 1 percent level.

**Significant at the 5 percent level.

*Significant at the 10 percent level.

Column (3) adds variables for whether the firm is an R&D-performing firm, and the R&D SEP of employment at the firm. The estimated coefficients for both of the firm-level R&D variables are positive. The estimated coefficient for SEP is smaller compared to column (2), but is still substantial at 0.132.

Finally, column (4) presents the estimates for the production-function regression model with establishment fixed effects. This model removes any unmeasured cross-sectional establishment factors related to SEP and output productivity that would bias the estimate of the effect of SEP on output. Using only within-establishment variation in the regression variables, the fixed effects model provides our strongest test of the relation between SEP of employment and output productivity at the establishment level. In column (4) the estimated coefficient on SEP is 0.055, considerably smaller than in column (3), and the estimated coefficient on average years of education of workers is diminished by even more in proportional terms. Given that measurement error produces smaller estimated coefficients in longitudinal regressions than in cross-sectional regressions (Freeman 1984), the smaller estimates in the fixed effects regression are not surprising, and provides additional indication of the presence of measurement error.

5.2.1   Methods for Addressing Measurement Error

The first method we use to address measurement error in SEP and average years of education of workers is to use a weighted regression, where we weight observations in our production-function regression by the square root of the *number of matched workers* at the establishment. Since the sampling error of both variables depend inversely on the square root of the number of matched workers, this weighting procedure gives more weight to establishments with more precise estimates of these variables and less weight to establishments with less precise estimates, and should thus provide a better estimate of the effect of SEP on output.

Table 5.4, column (1), presents the weighted regression estimates for the OLS model in the restricted sample. The estimated coefficient on SEP increases by a factor of 1.71 (= 0.226/0.132) compared to the unweighted regression in table 5.3, column (3). The estimated coefficient on average years of education of workers increases by a factor of 1.13 (= 0.043/0.038). Table 5.4, column (2), presents weighted regression estimates for the establishment fixed effects model. Compared to table 5.3, column (4), the estimated coefficient on SEP more than doubles from 0.055 to 0.121. The estimated coefficient on years of education of workers also doubles from 0.004 to 0.008.[12]

---

12. A potential issue with the weighted regression method is that if the effect of SEP is heterogeneous and larger in establishments with more employment and more matched workers, then the weighted regression estimate of SEP will reflect both heterogeneity in SEP related to establishment size, and reduced measurement error due to sampling.

Table 5.4   Establishment production function, methods for addressing sampling error in variables, manufacturing establishments (1992–2007)

Method for addressing sampling error in variables	Weighted regression		James-Stein shrinkage adjustment			
Establishment sample	Restricted		Restricted		Matched	
Regression model	OLS (1)	Fixed effects (2)	OLS (3)	Fixed effects (4)	OLS (5)	Fixed effects (6)
Scientists and engineers proportion of employment	0.226*** (0.021)	0.121*** (0.031)	0.437*** (0.034)	0.373*** (0.054)	0.154*** (0.020)	0.147*** (0.031)
R&D firm	0.057*** (0.002)		0.065*** (0.002)		0.080*** (0.002)	
R&D scientists and engineers proportion of employment at the firm	0.244*** (0.011)	0.092*** (0.011)	0.144*** (0.011)	0.048*** (0.011)	0.122*** (0.009)	0.056*** (0.011)
Standard factor inputs						
Ln(employment)	0.345*** (0.002)	0.381*** (0.003)	0.350*** (0.002)	0.393*** (0.003)	0.358*** (0.001)	0.388*** (0.002)
Ln(capital equipment)	0.066*** (0.001)	0.047*** (0.002)	0.058*** (0.001)	0.042*** (0.002)	0.061*** (0.001)	0.044*** (0.001)
Ln(capital structures)	0.021*** (0.001)	0.018*** (0.002)	0.019*** (0.001)	0.009*** (0.001)	0.015*** (0.000)	0.007*** (0.001)
Ln(materials)	0.481*** (0.001)	0.354*** (0.001)	0.470*** (0.001)	0.340*** (0.001)	0.447*** (0.001)	0.330*** (0.001)
Ln(energy)	0.098*** (0.001)	0.117*** (0.002)	0.105*** (0.001)	0.119*** (0.002)	0.112*** (0.001)	0.111*** (0.001)
Employer attributes						
Multiestablishment firm	0.033*** (0.003)		0.035*** (0.002)		0.061*** (0.002)	
Establishment age	0.005*** (0.000)		0.005*** (0.000)		0.005*** (0.000)	
Establishment age squared	−0.000*** (0.000)	−0.000*** (0.000)	−0.000*** (0.000)	−0.000*** (0.000)	−0.000*** (0.000)	−0.000*** (0.000)
Production worker share of employment	0.020*** (0.006)	0.120*** (0.008)	−0.008 (0.005)	0.096*** (0.007)	−0.055*** (0.003)	0.046*** (0.004)
Average years of education of workers	0.043*** (0.001)	0.008*** (0.002)	0.054*** (0.001)	0.005** (0.002)	0.026*** (0.001)	−0.002 (0.001)
Number of observations	215,800	215,800	215,800	215,800	506,800	506,800
Adjusted R^2	0.931	0.965	0.915	0.958	0.956	0.976

Notes: Dependent variable is Ln(gross output). See table 5.2 for the definition of the "matched" and "restricted" establishment samples. The weighted regression method weights observations by the square root of the number of matched workers at the establishment. The James-Stein shrinkage-adjustment method is applied to the two independent variables that are constructed from the matched worker sample, that is, the scientists and engineers proportion of employment, and the average years of education of workers. For description of the method, see the text and appendix B. The OLS models include fixed effects for year (1992–2007), industry (six-digit NAICS), and geographic region (metropolitan or micropolitan core-based statistical area [CBSA] as defined in 2009 by the U.S. Office of Management and Budget, or economic area as defined in 2004 by the U.S. Bureau of Economic Analysis). The fixed effects model includes fixed effects for establishment and year. Standard errors are shown in parentheses.

***Significant at the 1 percent level.

**Significant at the 5 percent level.

*Significant at the 10 percent level.

The second method we use to address measurement error is to apply a James-Stein-type "shrinkage" adjustment to the estimated SEP for an establishment, which pulls the estimate toward the mean SEP in the entire sample, depending on the variance of the estimated SEP. Building on Mairesse and Greenan (1999), in our method, described in appendix B and in Barth et al. (2017), we calculate the ratio of the variance of estimated SEP at an establishment to the observed variance of SEP across all establishments. A large ratio indicates that sampling error in estimated SEP is large relative to the variation in SEP across establishments. We use this variance ratio to adjust the estimated SEP at an establishment toward the mean SEP over all establishments. We replace each establishment's estimated SEP with a weighted average of its estimated SEP and the mean SEP over all establishments in the data. The weight given to estimated SEP is smaller if sampling error of estimated SEP is larger, and the weight given to mean SEP is commensurately larger. The same procedure is applied to adjust the estimated average years of education of workers at the establishment.

Columns (3) and (4) of table 5.4 present the regression estimates for the OLS model and establishment fixed effects model using the restricted sample, and applying the James-Stein shrinkage adjustment to SEP and average years of education of workers. In table 5.4, column (3), for the OLS model, the estimated coefficient on SEP is 0.437, which is 3.31 (= 0.437/0.132) times larger than the comparable estimate in table 5.3, column (3). The estimated coefficient on average years of education of workers is 0.054, which is 1.42 (= 0.054/0.038) times larger than the comparable estimate in table 5.3, column (3). For the establishment fixed effects model, in table 5.4, column (4), the estimated coefficient on SEP is 0.373, which is dramatically larger than the comparable estimate of 0.055 in table 5.3, column (4). The estimated coefficient on years of education of workers is 0.005, which is only marginally larger than the comparable estimate of 0.004 in table 5.3, column (4).

The last two columns in table 5.4 present regression estimates using the matched sample of all establishments, and applying the James-Stein shrinkage adjustment to SEP and average years of education of workers. The estimated coefficient on SEP in the OLS model is 0.154, which is almost double the comparable estimate of 0.079 in table 5.3, column (1). The estimated coefficient on SEP in the establishment fixed effects model is 0.147, which is very close to the OLS estimate of 0.154.

In sum, the regression estimates presented in tables 5.3 and 5.4 show that the different methods for addressing measurement error in the SEP variable all lead to larger estimates for the effect of SEP on output in the establishment production function. We conclude that the SEP of employment has a substantial positive impact on output productivity at the establishment in our data for manufacturing establishments in 1992–2007.

5.3 Establishment Scientists and Engineers and Earnings of Workers

If the SEP of employment is positively related to productivity at the establishment, then we may expect that it is also positively related to the earnings of workers at the establishment. A positive relation between SEP and worker earnings would result if new technologies implemented by scientists and engineers at the establishment complement worker skills in production,[13] or if employers share economic rents from implementing those technologies or products with workers through higher pay.

The standard earnings equation regression model in labor economics relates individual workers' log earnings to their human capital and demographic attributes. To assess the effect of SEP on worker earnings, we augment the standard earnings equation with SEP at the establishment and R&D SEP of employment at the firm. We estimate the following workers' earnings regression model:

$$(2) \quad \text{Ln EARN} = a + b\,\text{SEP} + c\,\text{FRD} + d\,\text{FRDP} + \textbf{WKR}\,\gamma + \textbf{EMPL}\,\delta$$
$$+ \textbf{YR} + \textbf{IND} + \textbf{GEO} + u,$$

where

EARN is annualized earnings of the worker;
SEP is the scientists and engineers proportion of employment at the establishment;
FRD is an indicator for whether the firm is an R&D performing firm;
FRDP is the R&D scientists and engineers proportion of employment at the firm;
WKR is a vector of individual worker attributes, including years of education, years of work experience, years of work experience squared, indicator for female, indicator for nonwhite race, indicator for scientist or engineer occupation, and interactions of indicator for female with work experience, work experience squared, and indicator for nonwhite race;
EMPL is a vector of other employer attributes, including log employment at the establishment, production worker share of employment at the establishment, and average years of education of workers at the establishment;
YR is a vector of year fixed effects;
IND is a vector of industry fixed effects;
GEO is a vector of geographic region fixed effects; and
u is the error term.

Table 5.5 describes our sample of workers in manufacturing establishments in 1992–2007. This sample of 11,666,200 person-year observations

---

13. Some technologies substitute for labor skills and reduce earnings, so complementarity of technology and labor skills depends on the specific case.

corresponds to the 215,800 establishment-year observations in the "restricted" establishment sample presented in table 5.2. The mean value of SEP in the worker sample, 0.063, is greater than the mean value of SEP in the establishment sample, 0.038, because larger establishments with more workers have higher SEP.

The last three columns in table 5.5 compare workers who in our panel have work history in R&D firms only, in both R&D and non-R&D firms, and in non-R&D firms only. Workers with work history in R&D firms only have higher earnings, more years of education, and work in establishments with more employees and higher SEP compared to workers with work history in both R&D and non-R&D firms, or non-R&D firms only.

Table 5.6 presents the regression estimates for the log earnings equation augmented by SEP and other establishment-level and firm-level variables. In column (1), using the matched sample of workers, the estimated coefficients on the human capital and demographic variables—years of education, years

Table 5.5   **Mean value of selected variables, workers in manufacturing establishments (1992–2007) (restricted sample of workers)**

		Workers with work history in		
	All workers	R&D firms only	Both R&D and non-R&D firms	Non-R&D firms only
Number of person-year observations	11,666,200	8,173,300	1,572,800	1,920,200
Worker attributes				
Ln(earnings)	10.42	10.50	10.31	10.18
Years of education	12.4	12.6	12.0	11.6
Years of work experience	23.9	24.1	22.9	23.7
Female	0.311	0.312	0.287	0.324
Nonwhite race	0.256	0.239	0.285	0.307
Scientist or engineer occupation	0.063	0.077	0.040	0.019
Employer attributes				
Ln(employment) at the establishment	6.21	6.50	5.83	5.30
Production worker share of employment at the establishment	0.714	0.699	0.744	0.755
Establishment in R&D firm	0.780	1	0.592	0
R&D scientists and engineers proportion of employment at the firm	0.045	0.059	0.026	0
Scientists and engineers proportion of employment at the establishment	0.063	0.077	0.040	0.020
Average years of education of workers at the establishment	12.3	12.6	12.0	11.5

Notes: The "restricted" sample of workers are all workers in the LEHD that match to the Decennial Census (1990, 2000) or Current Population Survey (1986–1997), and also match to the "restricted" sample of manufacturing establishments presented in table 5.2. Worker earnings, gender, race, and age are from the LEHD. Years of work experience is constructed from worker age and education. Worker occupation and education are from the Decennial Census or Current Population Survey. See table 5.2 for definition of scientist or engineer, and description of establishment-level and firm-level variables.

Table 5.6   **Workers earnings equation, workers in manufacturing establishments (1992–2007)**

Worker sample	Matched	Restricted	Restricted	Restricted
Regression model	OLS	OLS	Job stayers	Job changers
	(1)	(2)	(3)	(4)
Scientists and engineers proportion of	0.582***	0.638***	0.010**	0.184***
employment at the establishment	(0.003)	(0.004)	(0.005)	(0.010)
R&D firm	0.067***	0.057***		0.047***
	(0.000)	(0.000)		(0.001)
R&D scientists and engineers proportion	0.119***	0.100***	−0.007***	0.018***
of employment at the firm	(0.002)	(0.002)	(0.001)	(0.004)
Worker attributes				
Scientist or engineer occupation	0.150***	0.150***		
	(0.001)	(0.001)		
Years of education	0.060***	0.061***		
	(0.000)	(0.000)		
Years of work experience	0.045***	0.044***		
	(0.000)	(0.000)		
Female	−0.137***	−0.141***		
	(0.001)	(0.001)		
Nonwhite race	−0.156***	−0.148***		
	(0.000)	(0.000)		
Employer attributes				
Ln (employment) at the establishment	0.040***	0.036***	0.057***	0.022***
	(0.000)	(0.000)	(0.000)	(0.000)
Production worker share of	−0.042***	−0.019***	0.043***	0.022***
employment at the establishment	(0.001)	(0.001)	(0.001)	(0.002)
Years of education of workers at the	0.043***	0.057***	−0.003***	0.032***
establishment	(0.000)	(0.000)	(0.000)	(0.001)
Number of observations	12,966,900	11,666,200	10,263,700	1,656,200
Adjusted R^2	0.553	0.563	0.902	0.818

Notes: Dependent variable is Ln(earnings). The "matched" sample of workers are all workers in the LEHD that match to the Decennial Census (1990, 2000) or Current Population Survey (1986–1997), and also match to the "matched" sample of manufacturing establishments presented in table 5.2. The "restricted" sample of workers are all workers in the LEHD that match to the Decennial Census (1990, 2000) or Current Population Survey (1986–1997), and also match to the "restricted" sample of manufacturing establishments presented in table 5.2. The OLS models include fixed effects for year (1992–2007), industry (six-digit NAICS), and geographic region (metropolitan or micropolitan core-based statistical area [CBSA] as defined in 2009 by the U.S. Office of Management and Budget, or economic area as defined in 2004 by the U.S. Bureau of Economic Analysis). Job stayers observations are person-year observations for a worker at an establishment in two or more continuous years. The job stayers model includes fixed effects for person and year. Job changers observations are person-year observations for a worker before and after a change in the establishment. The job changers model includes fixed effects for person, year, industry, and geographic region. All models also include years of work experience squared and interactions of female with years of work experience, years of work experience squared, and nonwhite race. Standard errors are shown in parentheses.

***Significant at the 1 percent level.

**Significant at the 5 percent level.

*Significant at the 10 percent level.

of work experience, gender, and race—are similar to typical estimates in standard earnings equations. The estimated coefficient on scientist or engineer occupation is 0.150, so scientists or engineers earn 15 percent more than other workers. The estimated coefficient on SEP is a substantial 0.582. In column (2), using the restricted sample of workers, the estimated coefficient on SEP increases to 0.638, presumably due to reduced measurement error in SEP in the restricted sample. In the OLS model, the 0.638 estimated effect of SEP on earnings in table 5.6 is larger than the estimated effect of SEP on establishment productivity in tables 5.3 and 5.4, which ranges between 0.132 and 0.437, in the restricted sample.[14]

We consider the possibility that the relation between SEP and earnings may be affected by dual causality that produces a selectivity bias in the estimate. Establishments choose which workers to make job offers to, and workers choose which job offers to accept, so part of the positive association between SEP and earnings could be due to selectivity in employer and worker choices rather than the effect of science and engineering on the earnings of a given worker. The natural way to control for this selectivity is to estimate a fixed effects model that identifies the effect of changes in SEP on the same worker. In our data, there are two distinct ways that SEP can change for a given worker. An employer can change SEP over time, which affects workers who stay at the establishment, or a worker can move between establishments with different levels of SEP. Given the different impetus for change in these situations—an employer-initiated change versus a worker-initiated change—we estimate fixed effects models separately for job stayers and for job changers.

Table 5.6, column (3), presents estimated coefficients from a fixed effects analysis of *job stayers*, defined as person-year observations for a worker at an establishment in two or more continuous years, where changes in SEP are within the establishment over time. Table 5.6, column (4), presents estimated coefficients from a fixed effects analysis of *job changers*, defined as person-year observations for a worker before and after a change in establishment, where changes in SEP are due to the change in employer.

There is a large difference between the two fixed effects estimates for the coefficient on SEP. For job stayers the coefficient on SEP is a modest 0.010, while for job changers the estimated coefficient on SEP is 0.184. Workers benefit mainly by changing jobs and moving to establishments with higher SEP of employment, and not from their employer raising the SEP of employment at their current work establishment. Comparing the fixed effects estimate of the impact of SEP on earnings for job stayers at

---

14. Since labor share is around 0.35 to 0.40 in the gross output production function, an estimated effect of SEP on earnings that is greater than the estimated effect of SEP on gross output productivity is not inconsistent with rent-sharing of productivity gains.

an establishment with the fixed effects estimates of the impact of SEP on productivity at the establishment, we find that increasing the SEP at the establishment has a much greater impact on productivity than on earnings. From tables 5.3 and 5.4, the fixed effects estimate of the impact of SEP on productivity in the restricted sample ranges from 0.055 to 0.373, which are much larger than the 0.010 estimate of the impact of SEP on earnings of workers who are job stayers at an establishment.

5.4 Conclusion

Linking data on science and engineering occupations of workers, firm R&D activity, establishment production, worker earnings and job mobility, we find that goods-producing establishments with relatively many scientists and engineers have higher productivity and worker earnings than those with few scientists and engineers, and that the results hold up in fixed effects analyses that compare the productivity of the same establishment over time. A plausible interpretation of the results is that production-establishment-based scientists and engineers help implement the adoption of new technologies and products at workplaces. In addition, we find that earnings of workers are higher at establishments with higher proportions of scientists and engineers, but that the positive relation between earnings and SEP of employment is mainly due to workers moving to establishments with higher numbers of scientists and engineers rather than existing establishments increasing SEP. Our estimates of the effect of SEP of employment on productivity at the establishment are substantially strengthened when we apply methods to address measurement errors due to sampling variance in the estimate of the variable SEP.

Given that most industrial scientists and engineers work at goods- and services-producing establishments and that most of that work is not in formal R&D activities, our analysis suggests that there is much to be learned from extending studies of the economic effects of R&D on the economy to the effects of scientific and engineering work more broadly. Further study using qualitative as well as quantitative techniques could illuminate the link between R&D and non-R&D scientists and engineers and economic performance beyond our foray into this area: ethnographic studies of the work activities of production-establishment-based scientists and engineers compared to those of other high-level professionals in bringing new processes and products to the market, statistical analysis of nonmanufacturing industries where scientists and engineers increasingly play an important role in implementing information technology and other new technologies, and analysis of the endogenous decisions of firms to employ more or fewer scientists and engineers over time and to allocate them to R&D or non-R&D activities.

Appendix A

The Employer-Employee-Scientist-Engineer Data Set

This appendix describes how we link establishment-, person-, and firm-level data files to create the time series cross-section data set that we use in the chapter. We undertook this analysis at the Boston Federal Statistical Research Data Center (BRDC) where we followed all Census confidentiality requirements.

For our establishment data we use establishments from the Census of Manufactures (CM) and the Annual Survey of Manufactures (ASM) in 1992–2007. The CM provides quinquennial data for the universe of establishments. The ASM provides annual data for a sample of establishments in each year, including a certainty sample of establishments in large firms and a noncertainty sample of other establishments. We use data constructed by Foster, Grim, and Haltiwanger (2014) to measure the real value of output, capital stock for structures and equipment, and materials and energy use. We use the Fort and Klimek (2016) consistent six-digit NAICS industry coding of establishments to define the industry classification of establishments. We include only manufacturing establishments in our data set.

We link workers and employer reporting units observed in the Longitudinal Employer Household Dynamics (LEHD) database to establishments observed in the Longitudinal Business Database (LBD). This link utilizes matches at different levels of establishment, county, and industry detail developed in a crosswalk at the Census Bureau. We are thus able to link person-level data from the LEHD to establishment-level data from the LBD and CM/ASM. Our LEHD data are from thirty states with varying year coverage over 1992–2007.

We obtain data on workers within establishments from the LEHD Employment History Files (EHF) that provide quarterly data on the wages and jobs of individuals over time (Abowd et al. 2006). We use the EHF data to define a person's "main job" in a year as the job with the most quarters worked and highest annualized earnings. We exclude jobs with real annualized earnings less than the level of a full-time job at half the minimum wage in 2002. We limit our analysis to workers between eighteen and sixty-five years of age. The person-level data in the LEHD contains age, gender, and race, but does not include information on individual education and occupation.

To identify occupation and education of individual workers, we link workers in the LEHD to the Decennial Census in 1990 and 2000 and to the Current Population Survey (CPS) in 1986–1997. We use a person-level crosswalk developed by the Census Bureau to link persons in the Decennial Census long-form sample in 1990 and 2000 and persons in the Current

Population Survey (CPS) over 1986–1997 to persons in the LEHD over 1992–2007. We first match persons between the LEHD and the 2000 census. For those who do not match, we match to the 1990 census, and for those who still do not match, we match to the CPS in 1986–1997. Using this procedure, we are able to create a matched sample of persons in our LEHD data for thirty states with varying year coverage over 1992–2007.

This method identifies occupation and education for 17 percent of all workers in LEHD employer reporting units that link to CM/ASM establishments in 1992–2007. Since our match identifies education and occupation of persons in 1990 or 2000 from the Decennial Census, or in 1986–1997 from the CPS, while our LEHD persons data covers the years 1992–2007, our matched sample of persons will miss younger workers or new entrants to the labor market to some extent. Using the matched sample, we estimate the SEP of employment at the establishment and the average years of education of workers at the establishment.

Finally, to measure the extent to which the parent firm of establishments conducts R&D we use the Survey of Industrial Research and Development (SIRD). The SIRD is an annual survey of firms, including a certainty sample of large R&D firms and a noncertainty sample of other firms. We use the SIRD panel of firms in 1977–2007 to construct our measure of R&D scientists and engineers employment as a share of total employment at the firm level. For each firm in the SIRD panel, we fill in missing years of R&D scientists and engineers employment share with the nearest nonmissing year available for the firm. We link the firm-level SIRD data to the establishment-level CM/ASM data. For CM/ASM establishments in firms that do not appear in any year of the 1977–2007 SIRD panel, the firm-level R&D scientists and engineers employment share is set to zero. From analysis of SIRD sample firms over the period, we conclude that we have reasonable coverage of smaller R&D performing firms.

Our full sample of establishments observed in CM/ASM over 1992–2007 includes all establishments with total employment of five or more. Our regression sample of establishments includes all establishments with ten or more workers matched to Decennial/CPS, and with total employment of fifty or more. Our regression sample of workers comprises all persons working in their "main job" in LEHD employer reporting units that link to our regression sample of CM/ASM establishments.

Appendix B

Shrinkage Adjustment to Address Measurement Error in Establishment-Level Scientists and Engineers Proportion of Employment and Average Years of Education of Workers

Because we measure scientists and engineers proportion (SEP) of employment and average years of education (AYE) of workers at manufacturing establishments from person-level matched data that has an underlying match rate of 17 percent, these two variables are measured with considerable sampling error. The error is greater the fewer the workers we match at an establishment and increases for any given number of matches as the number of employees increases at an establishment. In our regression analysis, regardless of how we treat measurement error, or even if we ignore it, our estimates of the effect of SEP on productivity are positive. But the magnitudes of the estimates vary considerably with the way in which we address the measurement error. The most novel way that we deal with measurement error is through shrinkage adjustment of the match-based variables. Building on Mairesse and Greenan (1999), we calculate the ratio of the estimated variance of estimated SEP (or AYE) at each establishment, associated with the number of matches and total employment, to the observed variance in estimated SEP (or AYE) across establishments, and use this ratio to shrink estimates with relatively large sample error to their mean value in a James-Stein-type shrinkage estimator.

Let n_{jt} be the number of matches we obtain from Decennial Census or CPS data for an establishment j with N_{jt} employees at time t. Let x_{ijt} be the matched variable under consideration (SEP or AYE) for person i at establishment j at time t—either a binary variable indicating whether a matched worker is a scientist or engineer, or a continuous variable for the years of education of a worker. We estimate SEP (or AYE) at the establishment with the sample mean

$$X_{jt} = \Sigma x_{ijt}/n_{jt}.$$

Following Mairesse and Greenan (1999), we estimate the variance of X_{jt} at the establishment with

$$V_{jt} = (1 - n_{jt}/N_{jt})(1/n_{jt})V^*,$$

where n_{jt}/N_{jt} is the sampling probability by establishment at time t, and $(1 - n_{jt}/N_{jt})$ is the correction factor for the finite sample,[15] and $V^* =$

15. We consider X_{jt}^*, the true value of SEP (or AYE) at the establishment, to be the value we want to use in our analysis, and we consider deviations of observed X_{jt} from true X_{jt}^* to be measurement error in X_{jt}. As the number of matches n_{jt} approaches the number of employees N_{jt}, the measurement error disappears. Mairesse and Greenan (1999) calculate variances of

$\Sigma \Sigma \Sigma (x_{ijt} - X_{jt})^2$ is the within-establishment sample variance of x_{ijt}, pooling over all establishment-years.

We calculate the reliability of variable X_{jt} as

$$\lambda(n_{jt}, N_{jt}) = 1 - (1 - n_{jt}/N_{jt})(1/n_{jt})(V^*/V),$$

where V is the observed sample variance of X_{jt} over all establishment-years.[16] We note that λ varies with n_{jt} and N_{jt}; reliability increases with the number of matches at the establishment, whereas, for a given number of matches, reliability decreases with the employment size of the establishment.

James-Stein estimators (James and Stein 1961) shrink estimates based on small samples toward some appropriate global value because the small samples have high sampling errors. Stein (1956) proved that while this yields a biased estimator of a value, it can reduce the variance of a model that uses the estimator. The intuition for the effect can be found in the problem of predicting the future productivity of a baseball player who has no hits in four at bats on the first day of the season while players on average have batting averages around 0.280 (i.e., they get hits 28 percent of the time). Given the small sample of four at bats it would be unrealistic to estimate the player's future batting average as 0.000. A better prediction for the rest of the season would be to take a weighted average of the small sample and the overall batting average of ball players or some other global average, such as the player's lifetime average or average over hundreds of at bats in the previous season. This approach is related to empirical Bayesian methods (see Efron 2010, chapter 1).

In our case, a James-Stein-type shrinkage estimator of X_{jt} is given by

$$Z_{jt} = \lambda_{jt} X_{jt} + (1 - \lambda_{jt}) X,$$

where X is the global mean and $1 - \lambda_j$ is a shrinkage factor that diminishes the role of the observed value and pulls it toward the global mean. As the estimated variance of X_{jt} increases, the shrinkage factor approaches unity and the shrinkage estimator of X_{jt} approaches the global mean X.

---

variables assuming the matched sample is drawn with replacement, whereas our calculation treats the matched sample as being drawn without replacement and therefore adds a correction factor for the finite sample.

16. More generally, in our implementation, we compute V as the variance of residuals from an establishment regression model for X_{jt} that includes covariates. For our establishment fixed effects model, to account for covariance between person observations over time, our computation of V^* includes an additional term, that is, one minus the harmonic mean of k_j/K_j across establishments, where k_j is the average number of years that unique persons are observed at establishment j during the panel, and K_j is the number of years that establishment j is in the panel; and we compute V as the variance of residuals from an establishment regression model for X_{jt} that includes establishment fixed effects and no other covariates. See Barth et al. (2017) for details of the methodology.

References

Abowd, J. M., B. E. Stephens, L. Vilhuber, F. Andersson, K. L. McKinney, M. Roemer, and S. Woodcock. 2006. "The LEHD Infrastructure Files and the Creation of the Quarterly Workforce Indicators." In *Producer Dynamics: New Evidence from Micro Data*, edited by T. Dunne, J. Bradford Jensen, and M. J. Roberts, 149–230. Chicago: University of Chicago Press.

Barth, E., J. C. Davis, R. B. Freeman, and A. J. Wang. 2017. "Using Person-Level Data to Address Measurement Error in Establishment-Level Estimates." Working Paper.

Bureau of Labor Statistics, U.S. Department of Labor. *Labor Productivity and Costs*. LPC Tables and Charts. www.bls.gov/lpc/tables.htm.

———. *Occupational Employment Statistics*. OES Data. www.bls.gov/oes/tables .htm.

———. 2012. *Standard Occupational Classification*. 2010 SOC Crosswalks. Options for defining STEM occupations under the 2010 SOC. Attachment B: STEM definition options. August. www.bls.gov/soc.

Efron, B. 2010. *Large-Scale Inference: Empirical Bayes Methods for Estimation, Testing, and Prediction*. Institute of Mathematical Statistics Monographs. Cambridge: Cambridge University Press.

Fairman, K., L. Foster, C. J. Krizan, and I. Rucker. 2008. "An Analysis of Key Differences in Micro Data: Results from the Business List Comparison Project." Center for Economic Studies Paper no. CES-WP 08-28, Washington, D.C., U.S. Census Bureau. September.

Flood, S., M. King, S. Ruggles, and J. R. Warren. 2015. *Integrated Public Use Microdata Series, Current Population Survey: Version 4.0* (data set). Minneapolis: University of Minnesota. https://doi.org/10.18128/D030.V4.0.

Fort, T. C., and S. D. Klimek. 2016. "The Effects of Industry Classification Changes on U.S. Employment Composition." Working Paper, Dartmouth College, U.S. Census Bureau.

Foster, L., C. Grim, and J. Haltiwanger. 2014. "Reallocation in the Great Recession: Cleansing or Not?" NBER Working Paper no. 20427, Cambridge, MA.

Freeman, R. B. 1984. "Longitudinal Analyses of the Effects of Trade Unions." *Journal of Labor Economics* 2 (1): 1–26.

Griliches, Z. 1998. "Productivity and R&D at the Firm Level." In *R&D and Productivity: The Econometric Evidence*, edited by Z. Griliches, 100–133. Chicago: University of Chicago Press.

Hall, B. H. 2005. "Measuring the Returns to R&D: The Depreciation Problem." *Annales d'Économie et de Statistique* 79/80:341–81.

Hall, B. H., J. Mairesse, and P. Mohnen. 2009. "Measuring the Returns to R&D." NBER Working Paper no. 15622, Cambridge, MA.

Helper, S., and J. Kuan. 2016. "What Goes On under the Hood? How Engineers Innovate in the Automotive Supply Chain." NBER Working Paper no. 22552, Cambridge, MA.

James, W., and C. Stein. 1961. "Estimation with Quadratic Loss." *Proceedings of the Fourth Berkeley Symposium on Mathematical Statistics and Probability* 1:361–79.

Mairesse J., and N. Greenan. 1999. "Using Employee-Level Data in a Firm-Level Econometric Study." In *The Creation and Analysis of Employer-Employee Matched Data*, Contributions to Economic Analysis, vol. 241, edited by J. C. Haltiwanger, J. I. Lane, J. R. Spletzer, J. M. Theeuwes, and K. R. Troske, 489–512. Bingley, U.K.: Emerald Group Publishing Limited.

National Science Board. 2016. *Science and Engineering Indicators 2016*. Arlington,

VA: National Science Foundation (NSB-2016-1). www.nsf.gov/statistics/2016/nsb20161.

National Science Foundation, National Center for Science and Engineering Statistics. *Scientists and Engineers Statistical Data System (SESTAT)*. www.nsf.gov/statistics/sestat.

———. 2015. *Characteristics of Scientists and Engineers in the United States: 2013*. http://ncsesdata.nsf.gov/us-workforce/2013.

———. 2016. *Business R&D and Innovation: 2013*. Detailed Statistical Tables, NSF 16-313. Arlington, VA. www.nsf.gov/statistics/2016/nsf16313.

Ruggles, S., K. Genadek, R. Goeken, J. Grover, and M. Sobek. 2015. *Integrated Public Use Microdata Series: Version 6.0* (data set). Minneapolis: University of Minnesota. https://doi.org/10.18128/D010.V6.0.

Stein, C. 1956. "Inadmissibility of the Usual Estimator for the Mean of a Multivariate Normal Distribution." *Proceedings of the Fourth Berkeley Symposium on Mathematical Statistics and Probability* 1:197–206.

What Goes On under the Hood?
How Engineers Innovate in the Automotive Supply Chain

Susan Helper and Jennifer Kuan

6.1 Introduction

The questions addressed in this volume are motivated by the recognition that engineers play an important role in generating innovation and economic growth. In this chapter, we seek to offer some description of engineering work by looking in detail at a specific manufacturing industry—firms that supply automakers—to gain insight into how engineers create innovation. Autos account for 5 percent of gross domestic product (GDP) (International Trade Administration 2011), and have undergone significant innovation and improvement; over the period 1980 to 2004, average horsepower nearly doubled.[1] The auto sector has long been an important employer of engineers, with more than 80,000 engineers working in the sector in 2010.[2] At

Susan Helper is the Frank Tracy Carlton Professor of Economics at the Weatherhead School of Management at Case Western Reserve University and a research associate of the National Bureau of Economic Research. Jennifer Kuan is a visiting assistant professor at the Freeman School of Business at Tulane University.

For acknowledgments, sources of research support, and disclosure of the authors' material financial relationships, if any, please see http://www.nber.org/chapters/c12690.ack.

1. From 1980 to 2004, average horsepower for new passenger cars and light-duty trucks increased by 80 percent and 99 percent, respectively. Knittel (2012) estimates that fuel economy could have increased by 60 percent during this period had performance been held constant.

2. According to Occupational Employment Statistics from the Bureau of Labor Statistics, in 2010 there were 9,260 engineers in NAICS 3361 (motor vehicle assembly), 2,740 in NAICS 3362 (motor vehicle bodies), 34,040 in 3363 (motor vehicle parts), and 2,000 in 4231 (motor vehicle wholesale). As discussed below, NAICS 3363 does not do a good job of capturing firms in the motor vehicle parts sector. Using our survey data, we find that true employment in the motor vehicle parts sector is about twice that estimated for NAICS 3363 (White House Council of Economic Advisers 2013, chapter 7), so (assuming similar engineering intensity) we estimate about 68,000 engineers in motor vehicle parts. In addition, some engineers in the auto sector are employed as temporary help; they are not counted in the figures above. Adding these together gives our estimate of more than 80,000 engineers working in the auto sector in 2010.

the same time, the locus of engineering work has changed significantly. For example, in 2011, 70 percent of auto suppliers contributed design effort (a task typically performed by engineers) compared with 48 percent in 1989.[3] All of this makes the auto supply chain an important context in which to study engineering and innovation.

Our study also revisits themes from an earlier literature on incremental innovation that focused on manufacturing. Rosenberg (1963) describes nineteenth-century equipment makers who began inside manufacturing firms, but eventually spun out and helped spread technological change. Clark, Chew, and Fujimoto (1987) survey automakers about product development projects and find differences in the way firms from the United States, Japan, and Germany utilize suppliers when introducing new car models. Levin et al. (1987) take care to distinguish process innovation from product innovation in their survey of large industrial firms, and uncover different strategies for appropriating returns to innovation. All of this incremental improvement in processes and in product quality accumulates to produce economically significant change. Rosenberg and Steinmueller (2010, 15) examine engineering practices in aircraft and chemicals industries and point out that Douglas Aircraft's DC-3, which served 95 percent of U.S. air traffic, was "the product of innumerable small modifications and design improvements."

In the spring of 2011, we conducted a nationwide survey of thousands of firms in the supply chain. To design the survey, our research team performed dozens of detailed interviews, aiming for a broad picture of the industry. Thus, the plants varied by size, geographic location, and industry. We spoke with engineers, production workers, plant managers, sales managers, and human resources managers, and visited large and small firms, both in the Midwest, the traditional center of the auto industry, and the Southeast, an up-and-coming center of U.S. auto manufacturing. Industries also varied, including metalworking, assembly, chemicals, rubber, and electronics. Plant visits were indispensable in providing a sense of the type of innovative activities that engineers and others were engaged in, as well as the language for inquiring about such activity.

The size and strategies of supply chain companies vary tremendously. However, a majority of firms are small- to medium-sized, often family-owned, firms.[4] As we discuss below, most of these firms do not perform traditional research and development (R&D) or patent in the way that large firms do. This observation encouraged us to expand our definition of innovation to capture a more complete picture of innovative activity taking place

3. Data for 2011 comes from the survey described below; for 1989 data, see Helper (1994). In the earlier period, it was much more common for suppliers to produce parts designed entirely by their customers, the automakers.

4. Data from our survey suggests that firms with fewer than 500 employees account for about one-third of employment in the auto supply chain.

in manufacturing firms. We thus hope to broaden a more recent innovation literature that focuses on patent-intensive industries such as information and communications technology, biotechnology, and pharmaceuticals with additional measures of innovation. Our interviews also revealed the importance of customers for the innovative efforts of supplier firms. The suppliers we spoke with preferred certain Japanese automakers as customers because they shared expertise and helped suppliers improve. In contrast, our interviewees viewed their American customers as often making unreasonable demands for price reductions without offering much technical or organizational support. (See also Helper and Henderson 2014.)

The organization of this chapter is as follows: We first present an overview of our survey and respondent firms. Next, we describe the types of innovative activities performed by engineers and others inside those firms. This is followed by a discussion of the survey data about the workers engaged in this activity, and engineering tasks we measured. Finally, a brief description of customer effects concludes.

6.2 Overview of Survey and Respondent Firms

In order to survey the auto supply chain, we first had to identify firms in the supply chain, which extends several levels from automakers such as Toyota, GM, and Ford. Thus, one contribution of this study is to flesh out in a comprehensive fashion the reach of automotive manufacturing in the United States. We find an industry dominated by a few enormous firms, supported by thousands of small- and medium-sized enterprises (SMEs); our median respondent had about 100 employees. Another contribution of our survey is, therefore, a detailed view inside small manufacturing firms, for which little public data exist. In this section, we describe the process we used to identify firms in the auto supply chain and then provide a broad overview of those firms.

6.2.1 Identifying Auto Supply Chain Firms

A problem that has plagued research on the auto supply chain is that publicly available data do not provide a good picture of which establishments are currently in the auto supply chain. Many firms that supply the auto industry are not classified as auto parts manufacturers (3363, in the North American Industry Classification System, or NAICS), sometimes because they supply other industries and so do not self-identify as auto suppliers. At the same time, many firms than are in NAICS 3363 no longer supply the auto industry because managers of establishments bear the main responsibility for classifying themselves into NAICS codes and typically do not update these codes very often, even when their markets shift. Thus, in order to survey the automotive supply chain, we first had to determine which firms might be auto suppliers.

We assembled a list of candidate firms and establishments from eleven sources including ELM International, the Analyst Resource Center (ARC), the Michigan Manufacturing Technology Center (MMTC), the Original Equipment Suppliers Association (OESA), the Precision Metalforming Association (PMA), the Industrial Fasteners Institute (IFI), Ohio's Manufacturing Advocacy and Growth Network (MAGNET), the *Automotive News* Top 150 Suppliers list, Polymer Ohio, and the Michigan Automotive Research and Development Facilities Directory.

This last directory was particularly useful in identifying firms that specialize in automotive R&D because establishments performing R&D are classified in NAICS 54171, which at the most detailed category includes "R&D in the physical, engineering, and life sciences." Thus, it would be very difficult to extract from such a large class of firms those whose output is used primarily by the auto industry. A conservative count of establishments from the Michigan Automotive Research and Facilities Directory yields 25,000 employees in Michigan alone. A strictly NAICS-based analysis of the auto supply chain would fail to capture these highly skilled workers, and underestimate the employment, wages, and skill level of automotive production.

From the National Establishment Time Series (NETS) database, we selected firms that were in the auto supply chain using NAICS codes associated with the auto industry (C.A.R. 2010). In addition to 3363, these include functional specialties involved in auto manufacturing such as metal stamping, plastics manufacturing, and equipment producers, or those performing automotive-related R&D.

We used both manual and automated procedures to eliminate duplicate listings. Each of the firms was phoned and asked if they currently supply the auto industry. When called, over half of the establishments listed as NAICS 3363 said that they no longer supplied the auto industry; another one-third was out of business. Table 6.1 summarizes the outcome of this process. About 20 percent of our original list, or 3,800 firms, were likely automotive suppliers and only 37 percent of these firms were in NAICS 3363 ("motor vehicle parts manufacturing").

Three surveys were sent to each firm by email, web link, and mail, each one requiring different expertise: sales, plant management, and personnel. While the quality of responses received was likely to be high because respondents were likely to be knowledgeable about their particular area, the share of firms returning all three surveys was unfortunately low. Out of 1,411 responses (a response rate of 37 percent), only 98 returned all three surveys. Consequently, in our descriptive statistics, the number of respondents varies.

6.2.2   Description of Auto Supply Chain

The geographical distribution of respondent firms and likely auto suppliers (candidate firms on our list that did not respond to the survey) are similar. Figure 6.1 shows the locations of all plants in our sampling frame as dark

Table 6.1   **Construction of survey sample**

NAICS	Candidate firms by NAICS code	Percent
3363	Motor vehicle parts manufacturing	37.1
333514	Special die and tool, die set, jig, and fixture manufacturing	14.3
326199	All other plastics product manufacturing	12.8
332116	Metal stamping	1.4
332710	Machine shops	1.1
326220	Rubber and plastics hose and belting manufacturing	0.9
336211	Motor vehicle body manufacturing	0.2
	Other industries	32.3
	Total	**100.0**

N	Reason for elimination from sample	
3,646	Out of business	19.2
11,363	Not in auto industry	59.9
130	Duplicates	0.7
3,828	**Total remaining**	**20.2**

points and respondent plants as light points. The greatest concentration of auto supply chain firms is in Michigan, Ohio, and Indiana; similarly, nearly two-thirds of our respondents are from this tristate region.

While the survey was distributed to all firms identified as likely automotive suppliers, these firms fall into "tiers," with tier-1 firms supplying automakers directly and tier-2 firms supplying tier-1 suppliers, and so on. Some mega-suppliers, such as Visteon, Delphi, Magna, Lear, and Johnson Controls, have many billions of dollars in annual sales, almost all of which come from direct dealings with automakers. Tier-1 suppliers comprise just under 25 percent of our sample. Lower-tier firms tend to be smaller and more numerous. Figure 6.2 shows the distribution of respondents by number of employees. A majority of firms had fewer than fifty employees, and another 40 percent of respondents had fewer than 500. We estimate that about 30 percent of the automotive supply chain employment is at firms with fewer than 500 employees. About 40 percent of respondents were single-plant firms, and only 7.8 percent of respondents were unionized.

The average age of respondent firms was thirty-two years, with almost 60 percent of firms more than twenty-five years old. Figure 6.3 shows the age distribution of the 202 firms that gave their founding year. This suggests that while firms may not be entering the automotive business in large numbers, many incumbent firms are robust to the ups and downs of the auto industry. Some of the strategies used by firms to stay afloat resulted in little investment in human or physical capital; one-fourth of them had no engineers at all. Several firms we visited had very little debt, owning their land and equipment outright. These factors allowed them to survive as

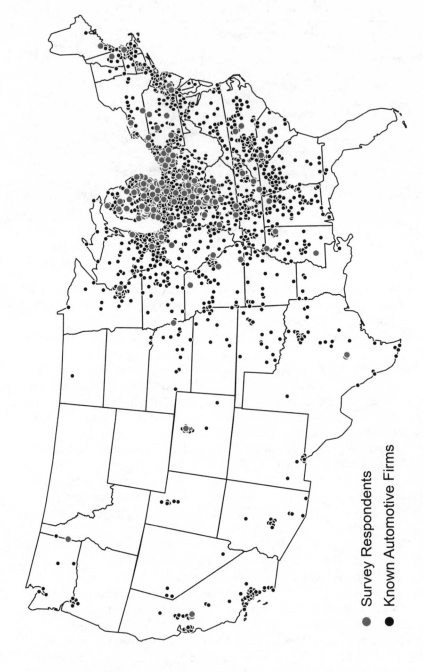

● Survey Respondents

● Known Automotive Firms

Fig. 6.1   Location of survey respondents

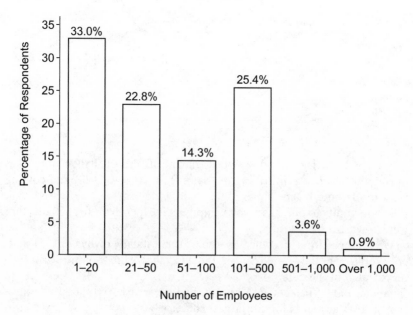

Fig. 6.2 Survey respondents by number of employees

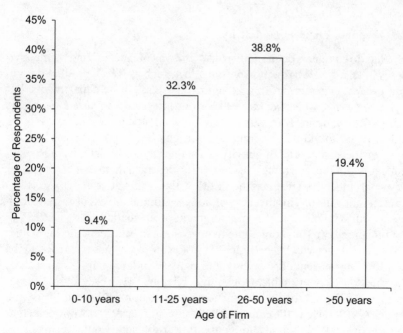

Fig. 6.3 Age distribution of supply chain firms

Table 6.2   Country of origin, customers, and suppliers

	U.S. customer	Japanese customer	German customer	Total	Percent of total
U.S. supplier	307	129	27	463	84.0
Japanese supplier	13	41	4	58	10.5
German supplier	14	7	9	30	5.5

"zombies" (one firm's self-description), laying off almost all their employees during the severe downturn in auto sales from 2009 to 2011, and coming back to life once business picked up again. One firm we interviewed took the opportunity to purchase used equipment inexpensively from plants that were downsizing or closing.

We surveyed foreign-owned as well as domestically owned auto suppliers operating in the United States. Table 6.2 shows the top three countries represented in the data. Most, 84 percent, are American, 10.5 percent are Japanese, and 5.5 percent are German. The survey asks only about the component that accounts for the largest share of automotive sales for a supplier; table 6.2 shows the country of origin of suppliers' main automotive customer.

6.3   Types of Engineering Activity

What do engineers do in small firms? One way of categorizing their activities is in terms of job function within a firm such as R&D, product design, and process engineering. In a large firm these might be departments, but in smaller firms, which we targeted for our interviews because most of the firms in our sampling frame are privately owned SMEs, the picture was quite different. We found that a company's only engineer might engage in one or more of these categories of activity in a single day. Some firms were so small that individuals performing these activities identified themselves as manager or owner. Indeed, some firms did not view their innovative activity as R&D or even innovation. This diversity of activities and fluidness of responsibilities within small firms makes measurement problematic.

Our survey can thus contribute to an existing literature on the challenges of measuring activity by job function. For example, in measuring R&D the literature has measured spending and patents, in large part because these data are publicly available. Cohen and Klepper (1996) review the literature on R&D spending, which focuses on large publicly traded firms, and Cohen (2010) reviews the empirical literature measuring innovation, including patenting. We inquired about patenting at smaller manufacturing firms. Even though patenting is an important measure of innovative activity or output, prior studies suggest that patents reflect an appropriability strategy, that is, more patents do not necessarily mean more invention, particularly when comparing across industries (Cohen 2010; Boldrin and Levine 2013). Our interviews and survey data were consistent with this notion of variable effectiveness of patents: certain types of firms engaged in patenting more than others, but overall, only 3 percent of our respondents applied for patents.

For product design, the literature has looked at new product introductions as a measure. This literature has the benefit of extending the large-firm analyses described above by including and even targeting smaller firms. Acs and Audretsch (1988, 1990) use product announcements to measure innovation by smaller firms, and Pavitt, Robson, and Townsend (1987) analyze British data on "significant" new products or processes, many of which are produced by small firms. Process engineering activity has tended not to be the focus of measurement, but a similar study by Leiponen (2005) examines manufacturing firms in Finland (many of which are SMEs) and finds that more skilled workers are more innovative. In our study, cost reductions are one measure of process engineers' efforts, particularly in a period of rising material costs.

Below, we discuss these categories of engineering activity, describing some context from our interviews and results from our respondents.

6.3.1 New Products

Measuring new product introductions can be an interesting indication of dynamism and change, even for a set of firms that sells products specified by their customer, because turning over products frequently requires flexibility and nimbleness. We asked, "What percent of your sales come from products that you did not make four years ago?"

At one factory we visited, an engineer had created a machine out of parts from two disused machines. The new machine was used to produce an item that had been produced by a Chinese competitor, but with higher quality and faster delivery.

Another low-tech solution to a customer's problem was found at a chemical company. While this rubber industry firm had several patents for rubber additives and had recently begun hiring chemists with doctoral degrees from a local university, one of its more profitable areas was its "cake-mix" product line. Managers at the firm had noticed their customers buying the same combinations of chemicals and having to measure and mix them. They had the idea to make easy-to-use, premixed packages, which would make their customers' outcomes more consistent and result in higher profits for the firm.

The introduction of new products designed by clever engineers is clearly a useful form of innovation. This does not mean that slow product turnover implies uninventive engineers. Rather, the frequent change in product line may be a distinct innovation strategy. Figure 6.4 shows the frequency of our responses. About a third introduced a new product at least every year, if not more.

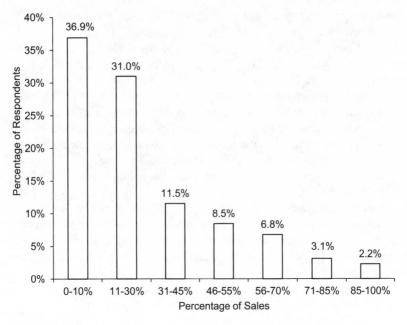

Fig. 6.4   What percentage of your sales comes from products that you did not make four years ago?

6.3.2   Design

In a manufacturing supply industry, engineers often take completed designs from their customers and produce them exactly as specified. In such cases, no additional design work is involved. On the other hand, the cake-mix products described above are designed entirely by supplier firm managers, albeit with considerable input from and observation of their customers. Because "design" is to some extent an innovative activity, we ask, "In the past year, roughly what percent of your plant's sales were from jobs where your firm designed the part or assembly?"

Figure 6.5 shows the considerable heterogeneity of responses to our question. A third of respondents fit the traditional model of supplier firm, producing only products designed by their customers. But about 15 percent produce parts that they design themselves. The rest, a majority of firms, fall somewhere in between these two extremes.

6.3.3   Innovative Contribution

Our broadest innovation question asks, "What percent of your sales come from products where you innovated in some way? By 'innovated,' we mean that your business unit designed a product with improved features compared to what the market had seen before, or that you used a novel process to make the product." (See figure 6.6.)

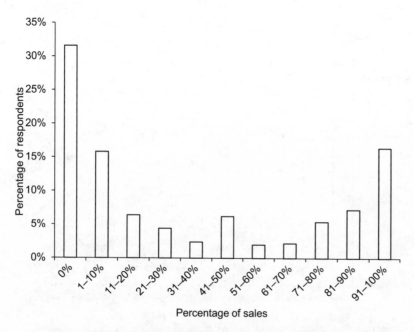

Fig. 6.5   In the past year, roughly what percentage of your plant's sales was from jobs where your firm designed the part or assembly?

This is a somewhat catchall question, but it is meant to encourage respondents to consider process innovations as well as product innovations when assessing their innovative contribution. One metal-stamping firm we visited was continually developing and extending its process capability, constantly creating new "know-how." For example, the owner of the firm worked with engineers to refine processes to accommodate very thin material, including plastics. This enabled the firm to serve an East Coast customer producing electric generators, a customer outside the usual geographical region and customer industry.

Even with our liberal definition of innovation, almost 42 percent of respondents report contributing little or no innovation to the products they make. At the other end of the spectrum, about 15 percent contributed some innovation to half their products or more.

6.3.4   Cost Reduction

Manufacturing cost reductions are often associated with economies of scale—as production quantities increase, the average cost goes down as a fixed component of cost is spread over more units. But process improvements can also generate cost reductions, for example, through fewer errors or more reliable equipment. We observed efforts in both of these directions on our plant visits.

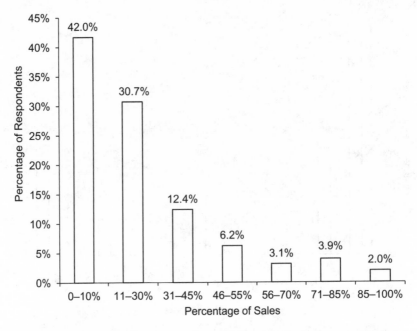

Fig. 6.6 **What percentage of your sales comes from products where you innovated in some way? By "innovated," we mean that your business unit designed a product with improved features compared to what the market had seen before, or that you used a novel process to make the product.**

There were also clever applications of existing processes to reduce the cost of a component. A process engineer could help its firm win new business by extending and applying its process capabilities. For instance, one metal stamper we visited bought a large press capable of stamping inch-thick material, which was much thicker than the sheet metal that the current machinery was capable of forming. This stamper was able to produce a part that had previously been produced by casting, a much more costly and energy-intensive process that involves pouring molten metal into a mold.

At another firm, process engineers added welding capability to the firm's production technologies. This allowed the firm to make a part using two stamped pieces that were joined together by welding. The welded assembly replaced a more costly cast part that its customer was importing from a low-wage country.

Figure 6.7 shows responses to our question about cost reduction. About 15 percent of firms reduced their costs, with another 40 percent maintaining cost levels. These responses are particularly interesting during our survey period because commodity prices were increasing. Thus cost reductions, and even holding costs level, are likely the result of successful engineering efforts. Note that at some firms, customers demand a schedule of price reductions. These price reductions might be accompanied by incremental process improvements, but if not, they could be met by a reduction in the firm's profits. Our question allows us to distinguish price reductions from cost reductions.

While these types of innovations produced by engineers at manufacturing firms seem minor, incremental innovations that reduce cost and create value for customers constitute a phenomenon of economic importance when aggregated across the thousands of firms that make up the auto industry. However, because each incremental innovation might seem unimportant, even to the engineer, measuring and valuing this activity can be difficult. Small firms rarely measure R&D spending, and their engineers tend to perform a variety of tasks, including innovation-related tasks. Many of the firms we interviewed eschewed the very term "innovation" as too sophisticated to describe their ongoing efforts to reduce cost and remain competitive. The experiments and development of new processes are carried out by the same engineers and technicians that maintain existing production lines and develop its traditional tooling, which helps explain why many firms lack careful formal accounting of R&D as a separate activity. We hope our additional measures can help overcome some of these issues for SMEs in manufacturing.

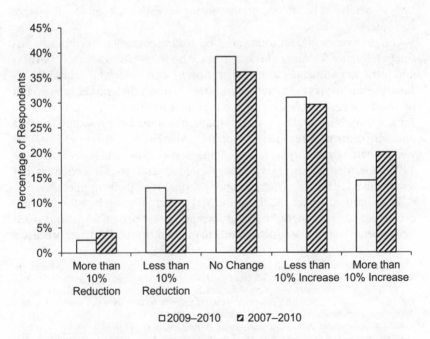

Fig. 6.7   **What has been the average annual percentage change in your unit costs?**

6.4   Engineering Interactivity

In the previous section, we categorized engineering activities in terms of function within a firm. In this section, we consider a different type of categorization: interactivity, with customers and with other functions inside the firm. We have described a work environment for manufacturing engineers of diverse activities, including maintaining existing processes on a daily basis but also developing new processes, refining and extending old processes, and solving customer problems to win new business. Amid this mix of activities, we also examine how and how much engineers interact with others. First we look at interaction with customers, using questions in our survey about certain investments and communications patterns. We then look at how much overlap there is between engineering tasks and the tasks of other types of workers, including skilled trades, production workers, and managers.

6.4.1   Interaction with Customers

The literature has long considered demand for a product to be relevant to innovation; Schmookler (1966) argued that innovation would be greater for goods that had a large market. In the context of a supply chain, demand considerations focus attention on buyer-supplier relationships, especially in the auto industry, with its oligopsonistic buyers. Outside of the corporate venture capital literature (e.g., Benson and Ziedonis 2010), relatively little research has been done on innovation by firms that sell to a few large customers.

Our interviews suggest wide variation among customers, including a distinct preference for Japanese customers, who were valued for their fairness and for their willingness to invest in suppliers. One steelmaker credited its Japanese customers for helping it improve so much that, rather than go out of business, it was able to compete even in a downturn. The literature on Japanese sourcing practices show Japanese automakers providing training and management assistance to suppliers (MacDuffie and Helper 1997), as well as their efficient organization of operations (Womack, Roos, and Jones 1990).[5] German customers were also viewed as fair by our interviewees, but less involved in the improvement and investment of their suppliers.[6]

To see how customer collaboration and communication benefited suppliers, we asked firms about "useful information that personnel at your plant have received on new products your firm might introduce and new processes

---

5. Some of this may be a result of practices in Japan where regular employees receive extensive on-the-job training, and employment norms have made it difficult for regular workers to change companies midcareer. This reduced labor mobility affects appropriability among Japanese firms, as company-trained employees are unlikely to take their skills with them to a competitor, or inventive employees take inventions to another firm.

6. As with Japanese customers, German customers might be affected by practices in their home country. There, automakers use small supplier firms aided by nationally subsidized training systems that produce highly skilled shop-floor employees who cooperate in R&D activities (Ezell and Atkinson 2011).

your firm might adopt." Figure 6.8 shows respondents obtaining both types of information, but slightly more product information than process information. Indeed, 80–90 percent of firms report getting some ideas from their customers.

We also asked about several specific engineering methods that we thought might be broadly representative of two different customer-service strategies: finite element analysis and value analysis/value engineering.

Finite element analysis (FEA) is the assessment of a component's suitability for its operating environment. Engineers use costly specialized software that incorporates scientific knowledge to evaluate an auto part's strength and durability in a given situation. For example, an engineer performing FEA on an engine component would use these software tools to judge whether the part was capable of withstanding the pressure, heat, impact, and other known environmental stresses it would be subject to, and whether the part could perform at the desired level of reliability and durability. The use of FEA tools requires that an engineer have specific training, as well as general scientific knowledge. However, this analysis can be performed as an independent task with a minimum of interaction with the customer.

By contrast, value analysis/value engineering (VAVE) involves extensive interaction between customer and supplier on a variety of design and manufacturing decisions. The purpose of VAVE is for suppliers to improve "value" to customers, which is defined as performance divided by cost.

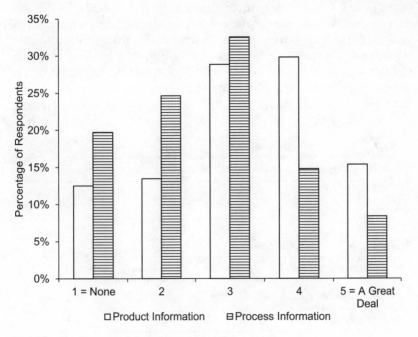

Fig. 6.8   Product and process information from customers

Engineers therefore make an effort to learn about their customer's needs broadly, and work with their customer to design a product or process. The chemical company that produced cake-mix products is one example of this type of customer-oriented approach, but efforts vary depending on the type of supplier firm and the extent of customer interaction. By contrast, a more conventional, non-VAVE approach to supplying components would take a customer's design as complete. The supplier would produce the part without modification or input. Many of our survey respondents take this more traditional approach.

Comparing the two, FEA involves an investment in equipment and engineers with specialized knowledge and skills, whereas VAVE requires engineers to spend time interacting with customers and to think broadly about customer problems and solutions. Thus, FEA and VAVE represent different strategies for investment and skills. However, they are not really polar opposites; some firms do both. Table 6.3 shows the breakdown of the 474 firms that responded to our questions about the use of FEA and VAVE, and we see that a majority of firms that provide VAVE services also provide FEA. Despite the proven effectiveness of both techniques, they remain rare in the U.S. industry; only one-third of respondents practiced VAVE, and only one-fourth had implemented FEA.

VAVE is only one measure of customer interaction. While a majority of firms, 60 percent of respondents, reported using neither FEA nor VAVE, some of these firms collaborate with customers outside of a VAVE framework. Table 6.4 shows responses to more general questions about customer interactions. We ask whether the supplier conducts regular or occasional visits with their customers, and more specifically, with their customers' engineers. The responses to both questions are almost identical; a majority of firms visit at least occasionally, with 30 percent visiting regularly. The engineering intensiveness of a firm's strategy is also reflected in employment data. Figure 6.9 shows the highly skewed distribution of engineering

Table 6.3   **Use of FEA and VAVE**

	VAVE (%)	No VAVE (%)
FEA	17	8
No FEA	15	59

Table 6.4   **Visits with customers**

	Customer (%)	Customer's engineers (%)
None	47	41
Occasional	22	28
Regular	31	31

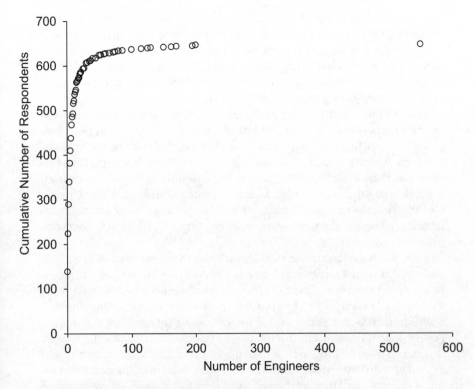

Fig. 6.9   Number of engineers employed by firms

employment at our respondent firms. Over 20 percent of the 647 firms that listed employment numbers had no engineers at all, and nearly one-third had just one to three engineers on staff. The picture that emerges is of a spectrum of firms ranging from low engineering intensity and low customer engagement to high engineering intensity and customer collaboration. Firms that perform VAVE had an average of 7.24 engineers on staff, compared with 4.76 for firms that do not perform VAVE. Similarly, firms that make regular visits with customer engineers had an average of 6.21 engineers compared with 4.06 for firms that do not make regular visits.

6.4.2   Interaction with Other Workers

One feature of Japanese management practice is "knowledge overlap" (Takeishi 2002; Helper and Sako 2010), whereby employees in one functional area gain an understanding of other job functions. This philosophy may seem to lead to duplication of effort, but the insight into how other jobs are performed improves employees' ability to communicate to solve problems and debug new operations. Thus, we measure engineers' interaction with other workers, especially skilled workers and unskilled production workers by asking whether nonengineers perform the following six engineering tasks: set up machines, modify programs on computerized equipment, diagnose equipment problems and inspect work in progress, use quality assurance data to recommend improvements, meet with customer personnel, and use a computer. Our idea is that the more "task overlap" there is, the more interaction nonengineering workers have with engineers.

Because one feature of Japanese manufacturing organization is close interaction between engineers and production workers, we compare task overlap at U.S. firms and Japanese firms. Table 6.5 shows, not surprisingly, much greater task overlap between skilled workers and engineers than unskilled workers and engineers. There is also a slightly higher rate of overlap for both types of workers within Japanese-owned firms in the United States. Out of six tasks we asked about, unskilled workers and engineers overlapped on 2.874 of them in Japanese-owned firms, and 2.25 in U.S.-owned firms.

Table 6.6 lists descriptive statistics for the variables used in table 6.7, and table 6.7 shows regression results for how overlapping of duties, or interaction between engineers and production workers, is associated with value-added per employee. Overlap with production workers is positive and significant. However, as the second column shows, this coefficient is no longer significant once we control for size, as represented by sales. This result is somewhat surprising because one might think that a larger size would allow more division of labor (rather than task overlap). The next two columns suggest that Japanese-owned firms translate knowledge overlap into productivity gains in ways that other firms do not; an interaction term for the knowledge overlap by Japanese firms is positive and significant, whether or not firm size is included. This result is consistent with the literature cited above, in that Japanese firms tend to place great importance on mechanisms for effective interaction between production workers and those above them in the hierarchy of the firm. Unfortunately, similar models of the effects of task overlap on our other innovation variables involved too few observations for good results.

Table 6.5   Number of tasks overlapping with engineers by U.S. and Japanese ownership

		U.S.	Japan
Unskilled/semiskilled	Average	2.25	2.74
	(Std. dev.)	(1.80)	(1.39)
	Range	0–6	0–6
Skilled	Average	4.51	4.81
	(Std. dev.)	(1.31)	(1.28)
	Range	0–6	0–6
N		404	31

Table 6.6   **Descriptive statistics**

Variable	N	Mean	Std. dev.	Range
R&D spending as a share of sales	449	3.19	1.83	1–7
Sales from new products	474	2.36	1.59	1–7
Sales from products containing your innovation	474	2.14	1.47	1–7
Cost reduction over past year	457	3.44	1.01	1–5
Cost reduction over past four years	453	3.50	1.06	1–5
Number of patents	1,431	0.30	5.24	0–142
Log (value added per employee)	211	4.23	0.97	−0.20–8.10
Used FEA	475	0.26	0.44	0–1
Used VAVE	474	0.32	0.47	0–1
Regular or occasional visits with customer engineers	265	0.90	0.84	0–2
Regular or occasional visits with customer	265	0.85	0.87	0–2
Overlap between engineers and unskilled workers	464	2.30	1.75	0–6
Overlap between engineers and skilled workers	464	4.53	1.31	0–6
U.S. supplier	1,431	0.75	0.43	0–1
Japanese supplier	1,431	0.06	0.23	0–1
German supplier	1,431	0.03	0.18	0–1
Other supplier	1,431	0.16	0.37	0–1

Table 6.7   **Value added per employee and engineering intensity, country and skill overlap (OLS)**

Dependent variable ln (value added)	(1)	(2)	(3)	(4)
Engineering intensity	0.61	1.04	0.64	0.92
	(1.09)	(1.02)	(01.07)	(0.99)
Japanese supplier	−0.27	−0.68	0.71	0.42
	(0.48)	(0.46)	(03.38)	(3.11)
German supplier	0.59	−0.41		
	(0.75)	(0.75)		
Other supplier	0.26	−0.09		
	(0.31)	(0.31)		
Unskilled overlap	0.13**	0.06	0.11*	0.04
	(0.06)	(0.06)	(0.06)	(0.06)
Skilled overlap	−0.02	−0.06	−0.03	−0.04
	(0.08)	(0.08)	(0.08)	(0.07)
Japanese * unskilled overlap			0.74*	0.75**
			(0.40)	(0.37)
Japanese * skilled overlap			−0.62	−0.64
			(0.75)	(0.69)
Size (ln sales)		0.23***		0.21***
		(0.06)		(0.05)
Constant	3.76***	2.11***	3.88***	2.19***
	(0.37)	(0.56)	(0.36)	(0.53)
R^2	0.06	0.20	0.09	0.23
N	94	94	94	94

***Significant at the 1 percent level.
**Significant at the 5 percent level.
*Significant at the 10 percent level.

Table 6.8   Correlation matrix

	R&D spending	New products	Innov. products	Cost red. (one year)	Cost red. (four years)	Patents	FEA	VAVE	Eng. visit	Cust. visit
R&D spending	1									
New products	0.06	1								
Innovated products	0.33***	0.21***	1							
Cost reduct. (1 year)	0.09**	-0.00	0.02	1						
Cost reduct. (4 years)	0.03	0.01	-0.02	0.62***	1					
Patents	0.07	-0.01	-0.03	0.00	-0.05	1				
FEA	0.16**	-0.06	0.11	0.09**	-0.05	0.08*	1			
VAVE	0.05	0.09*	0.07	-0.01	-0.00	-0.02	0.43***	1		
Eng. visit	0.29***	0.07	0.26***	0.01	-0.08	0.09	0.03***	0.35***	1	
Cust. visit	0.11*	0.01	0.12**	-0.02	-0.01	0.05	0.08	0.20***	0.51	1
U.S. supplier	0.00	0.06	0.04	-0.00	0.02	-0.01	-0.02	-0.17***	-0.14**	-0.06
Japanese supplier	-0.08*	-0.08*	-0.06	0.04	-0.01	-0.01	0.10**	0.06	0.11*	0.03
German supplier	0.08	0.01	0.03	0.04	0.06	0.10***	0.12***	0.06	0.06	0.05

Note: Correlation matrix including country variables ($N = 212$).

***Significant at the 1 percent level.

**Significant at the 5 percent level.

*Significant at the 10 percent level.

6.5   Discussion and Conclusion

An economically important industry that has produced significant gains in product performance, the auto industry increasingly relies on suppliers not only for manufacturing, but also for innovation. Engineers at supplier firms contribute innumerable incremental gains, many of which they themselves deem unworthy of the term "innovation." Nevertheless, skills and customer collaboration have generated a steady improvement in price and performance.

We hope our study contributes to the extensive literature on engineering and innovation by providing insight and detail about how engineers generate innovation, especially in a manufacturing context where patenting is uncommon. Our survey provides a variety of ways to measure innovative output, in addition to patenting and R&D spending. A correlation matrix of these measures shows how uncorrelated these additional measures are from the standard variables in the literature (table 6.8). Patenting, for example, is uncorrelated with R&D spending, new products, innovative products, or cost reduction. Employees with formal training as engineers contributed to most of these types of innovation; simple regressions of engineering intensity yielded positive and significant results for most of our innovation measures (productivity, R&D, innovative products), though not for new product introduction or cost reduction.[7]

Finally, our interviews serve to illustrate how diverse engineering activity can be. Engineers produce new chemical compounds, but also cake mixes; they build complex dies and stamping processes, but also cobble together two old machines. Employees without formal training as engineers participate in these engineering activities as well. The application of both formal training and on-the-job know-how seems to characterize firms that survive wide swings in demand and move technology forward.

References

Acs, Zoltan J., and David B. Audretsch. 1988. "Innovation in Large and Small Firms: An Empirical Analysis." *American Economic Review* 78 (4): 678–90.

———. 1990. *Innovation and Small Firms*. Cambridge, MA: MIT Press.

Benson, David, and Rosemarie Ziedonis. 2010. "Corporate Venture Capital and the Returns to Acquiring Portfolio Companies." *Journal of Financial Economics* 98 (3): 478–99.

Boldrin, Michele, and David K. Levine. 2013. "The Case against Patents." *Journal of Economic Perspectives* 27 (1): 3–22.

C.A.R. 2010. *Contribution of the Automotive Industry to the Economies of All Fifty States and the United States*. Ann Arbor, MI: Center for Automotive Research.

7. Results not shown.

Clark, Kim B., Bruce W. Chew, and Takahiro Fujimoto. 1987. "Product Development in the World Auto Industry." *Brookings Papers on Economic Activity* 3:729–81.

Cohen, Wesley M. 2010. "Empirical Studies of Innovative Activity and Performance." In *Handbook of the Economics of Innovation*, vol. 1 (Handbooks in Economics), edited by Bronwyn Hall and Nathan Rosenberg. Amsterdam: Elsevier.

Cohen, Wesley M., and Stephen Klepper. 1996. "A Reprise of Size and R&D." *Economic Journal* 106:925–51.

Ezell, Stephen J., and Robert D. Atkinson. 2011. *International Benchmarking of Countries' Policies and Programs Supporting SME Manufacturers*. Information Technology and Innovation Foundation. http://www.itif.org/files/2011-sme -manufacturing-tech-programss-new.pdf.

Helper, Susan R. 1994. "Three Steps Forward, Two Steps Back in Automotive Supplier Relations." *Technovation* 14 (10): 633–40.

Helper, Susan, and Rebecca Henderson. 2014. "Management Practices, Relational Contracts, and the Decline of General Motors." *Journal of Economic Perspectives* 28 (1): 49–72.

Helper, Susan, and Mari Sako. 2010. "Management Innovation in Supply Chain: Appreciating Chandler in the Twenty-First Century." *Industrial and Corporate Change* 19 (2): 399–429.

International Trade Administration, Office of Transportation and Machinery. 2011. *On the Road: U.S. Automotive Parts Industry Annual Assessment*. Washington, D.C.: United States Department of Commerce. http://www.trade.gov /mas/manufacturing/oaai/build/groups/public/@tg_oaai/documents/webcontent /tg_oaai_003748.pdf.

Knittel, Christopher R. 2012. "Automobiles on Steroids: Product Attribute Trade-Offs and Technological Progress in the Automobile Sector." *American Economic Review* 101:3368–99.

Leiponen, Aija. 2005. "Skills and Innovation." *International Journal of Industrial Organization* 23:303–23.

Levin, R. C., A. K. Klevorick, R. R. Nelson, and S. G. Winter. 1987. "Appropriating the Returns from Industrial Research and Development." *Brookings Papers on Economic Activity* 3:783–820.

MacDuffie, John Paul, and Susan Helper. 1997. "Creating Lean Suppliers: Diffusing Lean Production through the Supply Chain." *California Management Review* 39 (4): 118–51.

Pavitt, K., M. Robson, and J. Townsend. 1987. "The Size Distribution of Innovating Firms in the UK: 1945–1983." *Journal of Industrial Economics* 35:297–316.

Rosenberg, Nathan. 1963. "Technological Change in the Machine Tool Industry, 1840–1910." *Journal of Economic History* 23 (4): 414–43.

Rosenberg, Nathan, and Edward Steinmueller. 2010. "Engineering Knowledge." SIEPR Discussion Paper no. 11-022, Stanford Institute for Economic Policy Research, Stanford University.

Schmookler, J. 1966. *Invention and Economic Growth*. Cambridge, MA: Harvard University Press.

Takeishi, A. 2002. "Knowledge Partitioning in the Interfirm Decision of Labor: The Case of Automotive Product Development." *Organization Science* 13 (3): 321–38.

White House Council of Economic Advisers. 2013. *Economic Report of the President, 2013*. Washington, D.C.: U.S. Government Printing Office.

Womack, James P., Daniel Roos, and Daniel T. Jones. 1990. *The Machine that Changed the World: The Story of Lean Production*. New York: Scribner.

Engineers and Labor Markets

The Influence of Licensing
Engineers on Their Labor Market

Yoon Sun Hur, Morris M. Kleiner, and Yingchun Wang

Shortages and surpluses of engineers are a recurrent labor
market problem in the United States, which have attracted
considerable public and professional attention.
—R. B. Freeman (1976)

7.1 Introduction

Analysts of the labor market for engineers have often documented the
phenomenon of recurring booms and busts (Hansen 1961; Folk 1970; Freeman 1976). One potential public policy solution has been to regulate the
market for engineers, especially ones that require licenses to practice within
the occupation, and thereby reduce market volatility for engineers through
the use of occupational planners to determine the appropriate supply of
engineers. With this type of regulation and planning, perhaps these wide
swings, which result in uncertainty for both employers and those considering
entering the occupation, could be reduced. Besides the stated public policy
rationale that labor market regulation improves public health and safety, it
also may serve to reduce fluctuations in the market for engineers. Licensing
may create a "web of rules" that results in a more orderly functioning of
the labor market for the occupation that reduces uncertainty and variance
in quality and sets the rules for entry and exit within the occupation, which
provides a source of social insurance (Dunlop 1958). Further, engineers and

Yoon Sun Hur is an associate research fellow at the Korea Institute for International Economic Policy (KIEP). Morris M. Kleiner is professor of public affairs at the Humphrey School
at the University of Minnesota and a research associate of the National Bureau of Economic
Research. Yingchun Wang is assistant professor of management at the University of Houston,
Downtown.

We thank the editors for suggestions on earlier versions of the chapter. We appreciate the
work of Rebecca Furdek, William Stancil, and Timothy Teicher, and are grateful for their
excellent assistance with the statutory and other legal data used in this study. We also thank
Richard Freeman, Hwikwon Ham, Aaron Schwartz, and participants at the NBER/Sloan
Engineering Project Workshop for their helpful comments and advice. For acknowledgments,
sources of research support, and disclosure of the authors' material financial relationships, if
any, please see http://www.nber.org/chapters/c12688.ack.

the functioning of their labor markets are viewed as important contributors to innovation and economic growth. An analysis that sheds light on the functioning of these labor markets may contribute to an understanding of how institutional factors influence engineering's contribution to technological change. However, if the influence of licensing for engineers is similar to markets for other regulated occupations, it may then restrict the supply of labor, causing an increase in wages and a reduction in the utilization of engineers in production (Kleiner and Kudrle 2000; Kleiner and Todd 2009; Kleiner 2013).

The general policy issue of occupational licensing is an important and growing one in the U.S. labor market, since it is among the fastest-growing labor market institutions in the U.S. economy. For example, in the 1950s about 4.5 percent of the workforce was covered by licensing laws by state government (Kleiner 2006). By 2008 approximately 29 percent of the U.S. workforce had attained licensing by any level of government, and by the 1990s more than 800 occupations were licensed by at least one state (Brinegar and Schmitt 1992; Princeton Data Improvement Initiative [PDII] 2008; Kleiner and Krueger 2010, 2013). This figure compares with about 12.6 percent of the members of the workforce who said they were union members, another institution that looks after its members, in the Current Population Survey (CPS) for the same year; that value was down to 11.3 percent by the end of 2012 (Hirsch and Macpherson 2011; U.S. Department of Labor 2013). Although we do not have detailed information on the trends for the licensing of engineers, their level of unionization has declined, which is consistent with national trends. Figure 7.1 shows the decline in unionization for civil, electrical, and industrial engineers from 1983 to 2010. The steepest dip was for electrical engineers, where unionization declined from about 12.2 percent in 1983 to 4.8 percent in 2010. The smallest decline was for industrial engineers, whose rates of unionization declined from 9.2 percent to 8.3 percent over the same time period. We will focus our analysis on these engineering specialties for this chapter, since they represent a continuum of more to less regulated specialties in engineering.

Since occupational regulation has many forms, describing its various types is worthwhile. The occupational regulation of engineers in the United States generally takes three forms. The least restrictive form is registration, in which individuals file their names, addresses, and qualifications with a government agency before practicing their occupation. The registration process may include posting a bond or filing a fee. In contrast, certification permits any person to perform the relevant tasks, but the government—or sometimes a private, nonprofit agency—administers an examination or other method to determine qualifications and certifies those who have achieved the level of skill and knowledge for certification. For example, travel agents and car mechanics are generally certified but not licensed. The toughest form of regulation is licensure; this form of regulation is often referred to

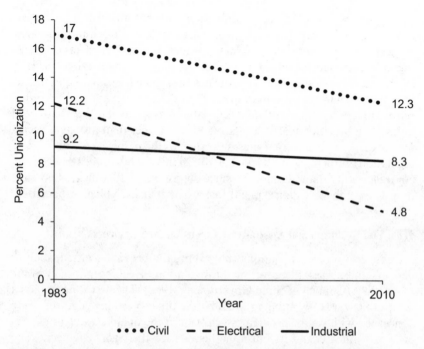

Fig. 7.1   Decline in unionization for civil, electrical, and industrial engineers, 1983–2010

Source: Current Population Survey, various years calculated by the authors.

as "the right to practice." Under licensure laws, working in an occupation for compensation without first meeting government standards is illegal. Our analysis provides a first look at the role of occupational licensing, rather than the other two forms of governmental regulation in the labor market for engineers in the United States.

We examine the role for occupational licensing in the labor market for engineers from 2001 through 2012. Initially, we present the evolution and anatomy of occupational licensing for engineers. Next, we present a theory of licensing and show how this form of regulation leads to wages dropping to the competitive wage as the licensing authority increases the supply of practitioners. In the following section, we show the data for the analysis and present the growth of regulation for the three types of engineers in our data set. Next, we present our empirical analysis for three large specialties in engineering—civil, electrical, and industrial—when variations in occupational licensing characteristics such as examinations and pass rates are included. In the final section, we summarize our results.

The theoretical model shows that government-granted licenses to protect the public can also lead to rents for the members of the occupation. As more individuals are allowed into the occupation by the planner, wages fall. To the extent that regulation reduces innovation and that unregulated members of the occupation can do higher wage tasks, regulation may diminish wages. The estimates in our models are small for the labor market effects of licensing, and they depend on the requirements and the engineering specialty examined. Also, some evidence indicates that some licensing requirements influence the number of hours worked by engineering specialty. The studies of the influence of licensing statutes on labor market outcomes perhaps need better data on individuals who have a license rather than just state licensing coverage, since coverage biases downward the influence of this type of regulation (Gittleman and Kleiner 2016). In this study we focus on licensing coverage rather than attainment, since determining attainment is possible only when individual data explicitly ask whether an individual was licensed.

7.2   The Evolution and Anatomy of Licensing for Engineers

Similar to other occupations that eventually became licensed, such as dentists and nurses, the government regulation of engineers began in the early 1900s (Council of State Governments 1952). The first state to pass a licensure law was Wyoming in 1907. At the time, Wyoming engineers were concerned with water speculators who lacked the qualifications or experience of trained engineers but nonetheless used the term "engineer." The law was passed so that "all the surveying and engineering pertaining to irrigation works should be properly done" (Russell and Stouffer 2003, 1). The American Society of Civil Engineers (ASCE) supported this piece of legislation, but otherwise resisted the notion of state-controlled licensing. After 1910, many civil engineering associations supported the concept of state licensing in order to control specific aspects of the practice that would be regulated. The ASCE promulgated a model law for licensure in 1910. This shift in policy also helped the occupation of civil engineering to be consistent with regulations that were being developed in other professions such as medicine and law, which had already accepted licensure (Haber 1991; Pfatteicher 1996). The effort to license the occupation was largely driven by engineering professional associations, but employers largely opposed rigorous regulations. Nevertheless, they were willing to agree to license the occupation if they also had the flexibly to employ unregulated practitioners to do most tasks.

Around 1920, the National Council of State Boards of Engineering Examiners was formed to work for licensure in every state, help enforce regulations, and ensure appropriate levels of experience and education for professional practice. This organization evolved into the National Council of Examiners for Engineering and Surveying (NCEES). As more states adopted regulations for professional practice, these engineering associations also became involved in advocating for the standardization of engineering curricula in professional schools and universities. It took nearly forty-five years for all fifty states to require licensure for the practice of civil engineer-

Table 7.1   **Percentage of engineers licensed by specialty, 1995 and 2012**

Engineering discipline	Percentage licensed 1995	Percentage licensed 2012
Civil	44	31
Electrical	9	9
Industrial	8	9

Sources: For 1995, Paul Taylor, *NCEES Licensure Bulletin*, December 1995; for 2012, the Survey of Income and Program Participation, 2013.

ing, although these licenses were required only for certain types of tasks that engineers perform.

In contrast, chemical, electrical, mechanical, and petroleum engineering were recognized as title holders and were covered by licensing following World War II. In the 1960s, industrial engineering was recognized as a title branch and was also regulated. Table 7.1 shows the percentage of engineers licensed by specialty in the United States, according to the National Council of Examiners for Engineering and Surveying (NCEES) in 1995 and from the Survey of Income and Program Participation (SIPP) for 2012. Civil engineering was by far the most regulated branch of engineering, with more than 44 percent of those practicing being licensed in 1995. This value declined to 31 percent in the SIPP in 2012. As the estimates in table 7.1 show, about 9 percent of the electrical engineers were licensed in the mid-1990s and in 2012, and about 8 to 9 percent of industrial engineers were licensed in the mid-1990s and in 2012. This suggests a large variance in the amount of regulation in the occupation of engineering. Moreover, the vast majority of engineers are covered by licensing statutes, but do not attain a license.

To measure the level of difficulty that each of the states sets for becoming a professional engineer, we develop an index of restrictiveness of engineer regulation. Not only has the level of licensing increased, but the intensity of the process of becoming licensed has become more difficult. Based on conversations with key officials at the NCEES, as well as with focus groups comprising engineers, architects, and interior designers, we have identified the following central items as important in becoming licensed: a general age/education requirement, experience requirements, a written exam, a practical performance exam, a specific engineering specialty exam, reciprocity requirements from other states, and a continuing education requirement.[1] These elements are the basis of an index of the rigor of the licensing process, in addition to the type of licensing. Using this index, we can trace the evolution of the intensity of the licensing index in the period 1995–2012. Table 7.3 summarizes the index of licensing regulations for engineers. The results show a slight upward movement in the mean values

1. We met and discussed with officials at the Minnesota Board of Architecture, Engineering, Land Surveying Landscape Architecture, Geoscience and Interior Design (AELSLAGID) regarding key criteria for licensing in that state and with several licensed engineers in Minnesota, Arizona, and California.

and a narrower spread in the variance of the licensing provisions across states. Occupational licensing is growing among states, and its provisions to enter and maintain good standing as a licensed professional engineer are becoming more stringent.

The nation's umbrella engineering-licensing body embraced a so-called Model Rule that would extend by thirty the number of extra credit hours engineers with bachelor of science degrees must attain to gain a professional license, but no state licensing board has made it a reality. However, the deadline for the professional association is in 2020. The goals of the licensing groups are to increase the status of engineers. For example, Blaine Leonard, former ASCE president and supporter of the increased requirements for becoming a licensed engineer, stated the following: "If we want to meet challenges and be prepared to protect the public, engineers need more depth of knowledge. You can't get it in programs under pressure." Proponents would like to see engineering attain the same professional status as medicine and accounting. The National Academy of Engineering and the National Society of Professional Engineers support the idea (Rubin and Tuchman 2012).

7.3 Basic Theory

To provide a theoretical context for our empirical work, we first review a model of the influence of licensing on the supply of labor. In the following section, we focus on the demand for labor and how government can be an important factor within a licensing model. The analysis of wage determination under licensing in engineering builds on work by Perloff (1980) on the influence of licensing laws on wage changes in the construction industry. The basic model posits that market forces are largely responsible for wage determination and that demand for work is highly cyclical. This approach also would apply to the engineering labor market. Perloff presents two cases. In the first, there are no costs to shifting across industries so that labor supply is completely elastic at the opportunity wage. In this case, the increase in the demand for work would have little effect on wages, since workers would flow between varying industries. The introduction of a licensing law renders the supply of labor inelastic. In this case, labor cannot flow between the sectors so that variations in demand would be reflected in the wage. In his empirical work, Perloff shows that for electricians, more so than for either laborers or plumbers, state regulations make the supply curve highly inelastic. Consequently, the ability of a state to limit entry or impose major costs on entry through licensing would enhance the occupation's ability to raise wages or allow them to fall during declines in the demand for labor, and retain levels of employment. We would expect that a similar approach would apply to the market for engineers, with more inelastic supply curves for civil engineers relative to electrical and industrial engineers.

Unlike the work that has been developed on the supply side, relatively little analysis has been done on how degrees of restriction of labor supply with occupational licensing influence wages and the amount of work, and how such restrictions of supply can make the labor market deviate from a competitive market. Our model focuses on the supply restriction of labor, and we develop a general model that we will apply to the regulation of engineers. We develop a model as follows:

Let $Q \equiv \Sigma_{i=1}^{n} q_i$, where q_i is each engineer's work output, n is the number of engineers in the market, and Q is the total quantity of supply. Each engineer's monetary utility function is $U_i = Q_i P(Q) - D_i(q_i)$, where D denotes the engineer's disutility and P is the price of work output (i.e., wage). The first-order condition for utility maximization is $P(Q) - D_i'(q_i) + q_i P'(Q) = 0$.

From the above equation, we have $[P(Q) - D_i'(q_i)]/P = (q_i/Q)/-[P/P'(Q)Q]$ $= (\alpha/\varepsilon)$ (1), where $\alpha \equiv q_i/Q$ is engineer i's market share, and $\varepsilon \equiv -[P'(Q)Q]$ is the elasticity of demand. Thus, the gap between price and marginal disutility is proportional to the engineer's market share and inversely proportional to the elasticity of demand. Price exceeds the engineer's marginal disutility as long as α is nonzero. The larger the difference, the more prices deviate from the socially efficient price.

For instance, for the symmetric case in which every engineer has the same output, with linear demand, $P(Q) = 1 - aQ$ for all i, and the convex disutility function being $D = bq + cq^2$. We assume that a, b, and c are greater than zero, so that the demand is inversely related to price and the disutility is a convex function. The first-order condition of the engineer's utility maximization becomes $1 - aQ - b - 2cq_i = 0$. The equilibrium is symmetric for this model: $Q = nq$, where q is the output per engineer. Hence, we obtain $q = (1 - b)/(an + 2c + a)$ (2). The market price is $p = b + [(1 - b)(2c + a)/(an + 2c + a)]$ (3), and each engineer's utility is $U = [(1 - b)^2(a + c)]/[(an + 2c + a)^2]$ (4).

The number of engineers, n, is an exogenous variable in the model. It is decided by the restrictions such as an examination requirement and the pass rate for the licensing exam. The stricter the licensing examination, the smaller the n. Equations (2) to (4) show that when the licensing requirement is stricter, the incumbent engineers' wage, output per person (measured as work hours in the empirical section), and utility will all increase.

From the above results and equation (1), we have $[(P(Q - D_i'(q_i)]/P = \alpha/\varepsilon = a(1 - b)/(abn + 2c + a)$. So another implication of the model is that the fewer the number of engineers, the further the wage will deviate from the socially efficient wage ($p = D'(q_i)$). Conversely, when the number of engineers becomes very large ($n \to \infty$), the wage tends to become the competitive wage. This would be the case when other unlicensed engineers can largely serve as substitutes for regulated engineers. Therefore, the prediction of the model is that the greater the supply restrictions, the larger the deviation from the competitive model.

Table 7.2   Key elements in development of the licensing index for engineers

Major components	Definition
Education requirement	Three if minimum level of education required to be licensed is bachelor's degree, two if it is associate's degree, one if board decides, otherwise zero
Experience requirement	Three if minimum level of education required to be licensed is eight years, two if it is four years, one if it is two years, zero if no requirement
Professional exam requirement	One if professional exam is required to be licensed, otherwise zero
Fundamental exam requirement	One if fundamental engineering exam is required, otherwise zero
Interim exam requirement	One if exam required for interim permit, otherwise zero
Continuing education requirement	One if state has any requirement for continuing education, otherwise zero
Specific exam requirement	One if specific additional exam is required for engineering discipline, otherwise zero

Source: Developed by the authors.

7.4   Data, Model, and Estimation

Using the above model as our guide, we now present the details of the information on the regulations facing engineers and the labor market conditions of the three broadly representative types of engineers: civil, electrical, and industrial. We chose these types of engineers because they reflect a continuum of regulation ranging from civil engineers who are the most regulated, electrical engineers less so, and industrial engineers the least regulated by state statutes. Table 7.2 displays the key elements (and their operational definition) of the licensing provisions in the statutes and administrative provisions that we plan to examine for each of the states in our sample for engineers.

Table 7.3 shows the yearly growth in the occupational licensing statutes index over the period 1995–2012. The results indicate that the occupation experienced growth in regulations governing entry and training requirements. The level of the index or the number of items included in the state-level measure grew from 6.94 to 7.25, or by about 4 percent over this time period. This reflects the intensity of the growth of requirements to enter and maintain the status as a licensed engineer. Further, the standard deviation declined by almost 23 percent, suggesting greater standardization of the requirements for licensing across states and over time.

Table 7.4 shows the relative ranking of the states that have the highest and lowest values in the index. We also developed values that were established through an expert systems focus group approach to test the sensitivity of the results to alternative methods of evaluation of standards. In this approach, an engineering student and a law student were given the data and

Table 7.3   **Growth of occupational licensing intensity over time**

Year	No. of state	Mean	Std. dev.	Min.	Max.
1995	51	6.94	2.04	0.00	9.00
1996	51	6.86	2.03	0.00	9.00
1997	51	6.89	1.86	0.00	9.00
1998	51	7.08	1.71	0.00	9.00
1999	51	7.08	1.71	0.00	9.00
2000	51	7.06	1.70	0.00	9.00
2001	51	7.06	1.70	0.00	9.00
2002	51	7.06	1.70	0.00	9.00
2003	51	7.06	1.70	0.00	9.00
2004	51	7.08	1.72	0.00	9.00
2005	51	7.08	1.72	0.00	9.00
2006	51	7.08	1.72	0.00	9.00
2007	51	7.08	1.72	1.00	9.00
2008	51	7.08	1.72	1.00	9.00
2009	51	7.08	1.72	1.00	9.00
2010	51	7.25	1.59	1.00	9.00
2011	51	7.25	1.59	1.00	9.00
2012	51	7.25	1.58	1.00	9.00

Note: Index is the summated rating value of the key provisions for licensing engineers as noted in table 7.2, tabulated by the authors.

Table 7.4   **Regulation rankings of top and bottom states by restrictiveness of licensing, 2009**

Top states	Index	Bottom states	Index
Pennsylvania	9	Virginia	1
Georgia	9	Minnesota	3
Texas	9	South Dakota	4
Illinois	9	District of Columbia	5
Arizona	9	Delaware	5
Colorado	9	Connecticut	5

asked to rank the states based on issues that were personally important to them as professionals in their respective fields. There was a high degree of consistency for the empirical and qualitative approaches. For many states, we were able to obtain the pass rates for the licensing examination for engineers.[2] Figure 7.2 shows the states and time trend in years for which we were able to obtain from the licensing boards of each of the states that posted their overall engineering pass rates.[3] The plots in the figure show

2. The links to the boards were available at http://ncees.org/licensing-boards. We went to the data from each state to obtain pass rates.
3. Tables 7A.1, 7A.2, and 7A.3 in the appendix show the influence of the variations in the pass rates and the influence of the statutory provisions of test requirements on both wages and hours of work for a limited number of states in which the data were available.

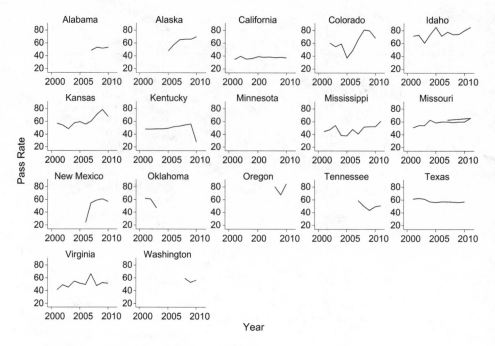

Fig. 7.2   Engineering exam pass rates by state

Sources: The pass-rate data for those states that post this information were obtained from http://ncees.org/licensing-boards/. In addition, we contacted the state boards to obtain pass rates for engineers for others not posted.

that California has the lowest steady-state pass rate for the engineering exam, averaging about 40 percent per year. In contrast, the pass rate for the licensing of engineers in Idaho is well above 80 percent. Unfortunately, no systematic national estimates could be developed because of the state data limitations over time and across states.

7.4.1   Economic Data

As a key part of our examination of the influence of regulation on the labor market for engineers, we use data from the American Community Survey (ACS) from 2001 through 2012. Table 7.5 presents the basic information that we used for our analysis. These variables include the standard variables from the ACS to include Mincer-type human capital variables such as gender, age, experience, education, and race. Unfortunately, no data on union status are available in the ACS. The means and standard deviations for the basic variables in the ACS are included in table 7.5 by type of engineer. They show that there are small differences in human capital characteristics such as age, experience, or education across engineering specialties. However, the percentage of civil engineers who work for the govern-

Table 7.5   Key variables for engineers in the ACS, 2001–2012

	Civil engineers		Electrical engineers		Industrial engineers	
	Mean	S.D.	Mean	S.D.	Mean	S.D.
Age	43.05	11.27	43.48	10.65	43.76	10.83
Schooling (in year)	16.00	1.67	16.21	1.66	15.70	1.66
Gender (male: 1; female: 0)	0.74	0.44	0.91	0.28	0.81	0.39
Married (married: 1; not married: 0)	0.73	0.45	0.76	0.43	0.74	0.44
Experience (in year)	21.05	11.40	21.28	10.89	22.06	11.11
Experience squared	572.88	495.33	571.25	473.94	610.37	492.57
White (white: 1; others: 0)	0.84	0.36	0.79	0.41	0.87	0.34
Black (black: 1; others: 0)	0.05	0.22	0.04	0.20	0.04	0.19
Citizen (U.S. citizen: 1; others: 0)	0.95	0.21	0.91	0.29	0.94	0.23
Work for for-profit (yes: 1; no: 0)	0.70	0.64	0.88	0.32	0.93	0.26
Work for not-for-profit (yes: 1; no: 0)	0.04	0.49	0.02	0.14	0.01	0.11
Work for government (yes: 1; no: 0)	0.24	0.62	0.09	0.28	0.05	0.22
Self-employment (yes: 1; no: 0)	0.05	0.49	0.01	0.12	0.01	0.08
Hourly earnings (in 2009 dollars)	34.61	21.20	37.47	18.41	30.35	14.72

Source: American Community Survey.

ment (about 24 percent) and are self-employed (about 5 percent) is much higher than in the other two types of engineering subgroups. The hourly earnings of electrical engineers (about $37 per hour) are the highest of the three categories. Generally, the licensing requirements for civil engineers have been in force the longest and are the most detailed across states. The estimates for hours worked are also derived from the ACS. Since there are more observations over time for civil engineers, we have information for all states and years for this category. For electrical and industrial engineers, however, some state and year observations are missing in the ACS, so states such as Wyoming, Hawaii, Montana, and South Dakota are missing observations for a couple of years in our sample.

7.4.2   Wage Determination

Our empirical strategy is to first examine the three categories of engineers—civil, electrical, and industrial—that may vary greatly by the type of regulation that influences their ability to find employment. We estimate the model using all engineers in the categories together and then estimate wage equations for each group separately. In figure 7.3 we show kernel density

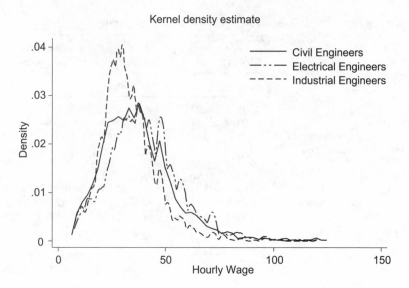

Fig. 7.3   **Empirical distribution of hourly rates for three types of engineers, 2009**
Source: Current Population Survey, various years.
Note: Sample includes those engineers who make above the minimum wage and excludes those with hourly wages greater than the top 1 percent.

plots for the three types of engineers in our study. The results show that electrical engineers have the highest mean value for wages and the widest distribution of earnings among the three types of engineers we study, but industrial engineers have the lowest mean value.

Our basic model uses an earnings function and compares the three types of engineers (the least regulated one—industrial engineers—is the excluded category). Our basic model is of the following form:

$$(1) \qquad \ln(\text{Earnings}_{ist}) = \alpha + \beta R_{ist} + \gamma X_{ist} + T_{ist} + \delta_{ist} + \theta_t + \varepsilon_{ist},$$

where Earnings_{ist} is the hourly earnings of engineer i at state s in year t; T_{ist} is the type of engineer (civil, electrical, or industrial) for person i's state s in year t; R_{ist} is the occupational licensing regulations (and its components) in person i's state s in year t; X_{ist} is the vector that includes covariates measuring characteristics of each person; δ and θ are state and year fixed effects, respectively; and ε_{ist} is the error term in our panel data.[4]

The model is a basic fixed effects approach that can also be viewed as

4. The use of a Rasch measure resulted in no basic effect of regulation on either wage determination or hours worked; consequently, we use the summated rating scale.

a generalization of the conventional two-group, two-period difference-in-difference model.[5] The estimates presented in the tables show the results for both a traditional panel estimate using individuals as the unit of observation of the role of regulation on wage determination and a two-stage estimation procedure. For the two-stage procedure, the first stage is developed by estimating a model of individual-level outcomes on covariates and a full set of state x (×) time fixed effects. The coefficients on the state x (×) time fixed effects represent state x time mean outcomes that have been purged of the variation associated with the within-cell variation in the covariates. In the second stage, these adjusted cell-level means are estimates on the policy variables and fixed effects. The two-step approach is a way of performing aggregation while still allowing for adjustment of individual-level covariates, which is a limitation of the pure aggregation. The basic panel estimates include individual covariates as well as state and year fixed effects.[6]

Table 7.6 shows estimates from the model developed from the overall licensing index on wage determination using both the individual observations and the two-level analyses with controls. Since the index is an imprecise measurement of regulation, we develop a relative measure of regulation of high, medium, and low levels of regulation using our index. We then compare the highest levels of regulation relative to the low and medium ones. The first column shows the basic bivariate relationship between having the most restrictive licensing statutes and wage determination with the full sample of the ACS. The basic relationship shows a statistically significant 2 percent effect.[7] However, in the second column, when human capital and state specific covariates are included, the estimates are still positive but small and not statistically significant. In examining the various engineering specializations in columns (3) through (8), we can see that there is some variation. For example, the bivariate estimates for civil engineers show a positive but small influence of being in a state with the most stringent regulations in the first stage, but no effect in the second stage. Similarly, for both electrical and industrial engineers, the engineering regulations have a small but positive effect in the first-stage bivariate estimates, but no influence in the second-stage results. The significant estimates range from a high of 4 percent with no covariates for industrial engineers with no covariates to no effect in the fully specified model. The categorical specifications show regulation for engineers has a small effect that is close to zero. For engineers, occupational licensing does not appear to influence wage determination for the occupation where

5. We also included time-varying state-level controls, such as the state median household income, but found that they have no explanatory power. Consequently, we do not show the results in this chapter.

6. The standard errors for these models were computed using a Huber-White covariance matrix that allowed for clustering at the state level.

7. We also estimated models that examined the influence of tougher licensing before and after the great recession of 2008, and found results similar to those presented in tables 7.6, 7.7, 7.9, and 7.10.

Table 7.6   **Influence of statutory rank index on wage determination: High relative to medium and low**

	(1) One-level analysis	(2) Two-level analysis	(3) One-level analysis	(4) Two-level analysis	(5) One-level analysis	(6) Two-level analysis	(7) One-level analysis	(8) Two-level analysis
Sample	All	All	Civil	Civil	Electrical	Electrical	Industrial	Industrial
Highest rank	0.024***	0.007	0.013***	−0.008	0.023***	0.026	0.042***	−0.014
	(0.000)	(0.016)	(0.001)	(0.023)	(0.001)	(0.028)	(0.001)	(0.034)
Observations	7,231,650	612	3,404,866	612	2,300,115	605	1,526,669	580
R^2	0.000	0.852	0.000	0.715	0.000	0.706	0.000	0.580
Basic control	No	Yes	No	Yes	No	Yes	No	Yes
Year fixed	No	Yes	No	Yes	No	Yes	No	Yes
State fixed	No	Yes	No	Yes	No	Yes	No	Yes

Note: Estimated with age, schooling in years, gender, marital status, experience, experience squared, race, U.S. citizenship, for-profit sector, and self-employment. Two-stage regressions are weighted by the number of engineers. The second-stage estimates are aggregate state-level estimates of hours worked calculated from the predicted hours worked in the individual model, which are then aggregated to the state level. The ACS sample uses individuals who earn less than $250 per hour and who are college graduates. Standard errors are in parentheses.

***Significant at the 1 percent level.
**Significant at the 5 percent level.
*Significant at the 10 percent level.

there is a relatively modest form of regulation with many unlicensed competitors. This is not unlike some of the specifications of the influence of unions on wage determination for other professional organizations (Lewis 1986). Moreover, we only have estimates of licensing coverage and not those who have attained a license, which may bias our results downward (Gittleman and Kleiner 2016). However, licensing requirements may also have effects on the supply of hours to the market.

The models developed for hours of work use approaches similar to the ones developed for our wage equation models. In a similar manner, we examine employment growth for each of the categories of engineers from 2001 to 2012. The basic model is of the following form:

$$\ln(E_{ist}) = \alpha + \beta R_{st} + \phi T_{ist} + \gamma X_{ist} + \delta_s + \theta_t + \varepsilon_{ist},$$

where Employment$_{ist}$ is hours of work per week per engineer in state s in time period t for individual i; T_{ist} is the type of engineer at state s in time period t for individual i; R_{st} is the regulation measure and its components at state s in time period t; the vector X_{st} includes covariates measuring economic and human capital characteristics within each state; δ_s and θ_t are state and year fixed effects, respectively; and ε_{ist} is the error term.

Table 7.7 gives the basic results for the effect of the licensing index on hours of work supplied by engineers using the specified model with regulations measured as in the high category relative to low or medium categorical measures of regulation. The results are consistent in showing that regulation is associated with an increase in hours worked by about 1 percent in the

Table 7.7   **Influence of statutory rank index on hours worked: High relative to medium and low**

	(1) One-level analysis	(2) Two-level analysis	(3) One-level analysis	(4) Two-level analysis	(5) One-level analysis	(6) Two-level analysis	(7) One-level analysis	(8) Two-level analysis
Sample	All	All	Civil	Civil	Electrical	Electrical	Industrial	Industrial
Highest rank	0.010***	0.009	0.010***	0.010	0.015***	0.012	0.005***	0.008
	(0.000)	(0.006)	(0.000)	(0.008)	(0.000)	(0.009)	(0.000)	(0.011)
Observations	7,231,650	612	3,404,866	612	2,300,115	605	1,526,669	580
R^2	0.001	0.335	0.001	0.202	0.002	0.216	0.000	0.196
Basic control	No	Yes	No	Yes	No	Yes	No	Yes
Year fixed	No	Yes	No	Yes	No	Yes	No	Yes
State fixed	No	Yes	No	Yes	No	Yes	No	Yes

Note: Estimated with age, schooling in years, gender, marital status, experience, experience squared, race, U.S. citizenship, for-profit sector, and self-employment. Two-stage regressions are weighted by the number of engineers. The second-stage estimates are aggregate state-level estimates of hours worked calculated from the predicted hours worked in the individual model, which are then aggregated to the state level. The ACS sample uses individuals who earn less than $250 per hour and who are college graduates. Standard errors are in parentheses.

***Significant at the 1 percent level.
**Significant at the 5 percent level.
*Significant at the 10 percent level.

bivariate estimates, but no effect when the two-level analyses is implemented with standard human capital controls. If regulation is effective in restricting the supply of new entrants to some extent, then those in the occupation are likely to work more hours. The results in table 7.7 are consistent with this hypothesis, but the magnitudes are small and significant in the one-stage estimates and insignificant in the two-stage results.

Although the categorical transformation of the overall index does not show much effect on the key labor market variables of wages and hours worked, perhaps several of the individual components of the licensing index may influence wages and hours worked. The use of an examination to determine the effect of this variable on wage determination has been used in other studies (Kleiner and Kudrle 2000; Kleiner and Krueger 2013). Through the examination process and the establishment of higher standards, access to and supply of engineers can be reduced, and if demand remains constant, wages can increase. Moreover, the pass rate for the engineering exam also may limit the entry of new engineers and drive up wages for engineers.

In table 7.8 we list the states that require a professional exam for each specific type of licensing examination. In order to become licensed, engineers usually take a fundamental or first exam, the basic step toward becoming a licensed engineer. This exam is often administered to engineers just prior to their finishing undergraduate studies. The professional exam, in contrast, covers general engineering practices and is usually given after engineers have been practicing for four or more years. It is the final stage of licensing coverage for entry into the regulated part of the occupation. Table 7.8 shows that Ohio and Arkansas adopted a professional exam in 2002 and 2009,

Table 7.8   **State professional exam requirements for licensure of engineers, 2001–2012**

Professional exam required	No professional exam required	Changer (year of change)
Alabama	Hawaii	Ohio (2002)
Alaska	Missouri	Arkansas (2009
Arizona	New Hampshire	
California	New Jersey	
Colorado	New Mexico	
Connecticut	Oregon	
Delaware	South Dakota	
District of Columbia	Utah	
Florida	Virginia	
Georgia	Washington	
Idaho	Wisconsin	
Illinois	Wyoming	
Indiana		
Iowa		
Kansas		
Kentucky		
Louisiana		
Maine		
Maryland		
Massachusetts		
Michigan		
Minnesota		
Mississippi		
Montana		
Nebraska		
Nevada		
New York		
North Carolina		
North Dakota		
Oklahoma		
Pennsylvania		
Rhode Island		
South Carolina		
Tennessee		
Texas		
Vermont		
West Virginia		

respectively; they serve as the basis for a difference-in-difference analysis. The difference-in-difference model is relative to Ohio and Arkansas, which were the states that changed their regulatory statutes for exams over the time period for which we have data and part of our analysis.

In order to provide sensitivity analysis for our previous estimates and include the estimates for an additional regulatory requirement, we include whether there is a professional exam requirement to become licensed.

Table 7.9 shows the estimates on wage and hours using seemingly unrelated regression (SUR) methods for the influence of having a professional exam requirement as part of the licensing requirement. Since only two states changed the exam requirements during the period under study, we used this method as an additional sensitivity test of our estimates. Panel A shows the results when engineers are categorized by type of engineering field: civil, electrical, and industrial. In panel B we estimate the model for all the engineers in our sample. Those estimates are consistent with the general results shown in tables 7.6, 7.7, and 7.8 and show only a small coefficient size for this requirement and varying levels of significance based on the type of engineering specialty and the labor market outcome variable selected, which was hourly wages or hours worked.

In table 7.10, we examine whether the lagged professional exam requirement variable may have influenced economic factors. Using the lagged professional licensing requirement and current economic data starting in 2001 through 2012, the table shows that these results are consistent in displaying a mixed to minor influence on wage determination. At least for licensing coverage, which is what our data allow us to measure, occupational licensing has a small influence on wage determination for civil engineers, but has a mixed influence on other engineering specialties. This may reflect the fact that the attainment of a license matters more with respect to wage determination, rather than the passage of a law regulating an occupation that is largely unregulated by the government (Gittleman and Kleiner 2016). Even though H. Gregg Lewis finds that being represented by a union raises wages by about 15 percent in aggregate, for many occupations such as hospital workers and well-educated male workers, the influence of unions is either zero or even slightly negative (Lewis 1986). Similarly, for civil engineers, who are more heavily licensed, tougher regulations may not enhance their earnings, perhaps because unregulated workers are able to be more innovative and create new markets relative to engineers who have their work standardized by the government (Friedman 1962).

7.5 Conclusions

Our chapter presents the first comprehensive analysis of the role of occupational licensing requirements on the labor market for civil, electrical, and industrial engineers. These groups of engineers represent among the largest number of engineers that are covered by occupational licensing statutes in the United States. We initially trace the historical evolution of licensing for engineers. Second, we present a theoretical rationale for the role of government in the labor market for the occupation. In the model, the government's ability to control supply through licensing restrictions and the pass rate limits the number of engineers, which may drive up wages. These results are useful for informing the empirical models for engineers.

Table 7.9 The influence of a professional exam on wages and hours worked for selected engineers

Panel A. Effect of a professional exam on wage and hours worked using seemingly unrelated regressions (SUR)

	Civil engineers				Electrical engineers				Industrial engineers			
	One-step		Two-step		One-step		Two-step		One-step		Two-step	
	(1)	(2)	(3)	(4)	(5)	(6)	(7)	(8)	(9)	(10)	(11)	(12)
	Log of wage	Log of hours worked	Log of wage	Log of hours worked	Log of wage	Log of hours worked	Log of wage	Log of hours worked	Log of wage	Log of hours worked	Log of wage	Hours worked
Index_pe	-0.002***	0.003***	-0.146***	-0.057***	0.003***	0.006***	0.042	0.012	-0.025***	-0.001**	-0.143**	-0.002
	(0.001)	(0.000)	(0.048)	(0.017)	(0.001)	(0.000)	(0.060)	(0.020)	(0.001)	(0.000)	(0.071)	(0.023)
Constant	3.437***	3.756***	3.637***	3.796***	3.520***	3.755***	3.537***	3.730***	3.351***	3.781***	3.551***	3.802***
	(0.001)	(0.000)	(0.056)	(0.020)	(0.001)	(0.000)	(0.068)	(0.023)	(0.001)	(0.000)	(0.081)	(0.027)
Observations	3,404,866	3,404,866	612	612	2,300,115	2,300,115	605	605	1,526,669	1,526,669	580	580
R^2	0.000	0.000	0.719	0.219	0.000	0.000	0.706	0.209	0.000	0.000	0.583	0.168
Basic control	No	No	Yes	Yes	No	No	Yes	Yes	No	No	Yes	Yes
Year fixed	No	No	Yes	Yes	No	No	Yes	Yes	No	No	Yes	Yes
State fixed	No	No	Yes	Yes	No	No	Yes	Yes	No	No	Yes	Yes

Panel B. Effect on wage using SUR for all engineers

	One-Step Analysis		Two-Step Analysis	
Variables	(1) Log of wage	(2) Log of hours work	(3) Log of wage	(4) Log of hours work
Index_pe	-0.006***	0.004***	-0.107***	-0.022*
	(0.000)	(0.000)	(0.034)	(0.012)
Constant	3.446***	3.760***	3.571***	3.775***
	(0.000)	(0.000)	(0.039)	(0.013)
Observations	7,231,650	7,231,650	612	612
R^2	0.000	0.000	0.854	0.303
Basic control	No	No	Yes	Yes
Year fixed	No	No	Yes	Yes
State fixed	No	No	Yes	Yes

Notes: Panels A and B were estimated with age, schooling in years, gender, marital status, experience, experience squared, race, U.S. citizenship, for-profit sector, and self-employment. Two-stage regressions are weighted by the number of engineers. The second-stage estimates are aggregate state-level estimates of hours worked calculated from the predicted hours worked in the individual model, which are then aggregated to the state level. The ACS sample uses individuals who earn less than $250 per hour and who are college graduates. Standard errors are in parentheses.

***Significant at the 1 percent level.

**Significant at the 5 percent level.

*Significant at the 10 percent level.

Table 7.10   Influence of lagged professional exam on wages and hours worked, 2001–2012

	All engineers		Civil engineers		Electrical engineers		Industrial engineers	
	(1)	(2)	(3)	(4)	(5)	(6)	(7)	(8)
Analysis	One-step	Two-step	One-step	Two-step	One-step	Two-step	One-step	Two-step
Panel A. Dependent is log of hourly wage								
Lag professional exam	0.000	−0.076**	0.003***	−0.066	0.008***	−0.004	−0.016***	−0.128**
	(0.000)	(0.030)	(0.001)	(0.043)	(0.001)	(0.053)	(0.001)	(0.063)
Constant	3.440***	3.564***	3.433***	3.558***	3.515***	3.589***	3.343***	3.493***
	(0.000)	(0.036)	(0.001)	(0.052)	(0.001)	(0.064)	(0.001)	(0.076)
Observations	7,231,650	612	3,404,866	612	2,300,115	605	1,526,669	580
R^2	0.000	0.853	0.000	0.716	0.000	0.706	0.000	0.583
Panel B. Dependent is log hours worked								
Lag professional exam	0.004***	−0.011	0.003***	−0.037**	0.007***	0.024	−0.001***	−0.010
	(0.000)	(0.010)	(0.000)	(0.016)	(0.000)	(0.018)	(0.000)	(0.021)
Constant	3.761***	3.763***	3.756***	3.772***	3.754***	3.717***	3.782***	3.807***
	(0.000)	(0.012)	(0.000)	(0.019)	(0.000)	(0.021)	(0.000)	(0.025)
Observations	7,231,650	612	3,404,866	612	2,300,115	605	1,526,669	580
R^2	0.000	0.300	0.000	0.213	0.000	0.211	0.000	0.168
Basic control	No	Yes	No	Yes	No	Yes	No	Yes
Year fixed	No	Yes	No	Yes	No	Yes	No	Yes
State fixed	No	Yes	No	Yes	No	Yes	No	Yes

Note: Estimated with age, schooling in years, gender, marital status, experience, experience squared, race, U.S. citizenship, for-profit sector, and self-employment. Two-stage regressions are weighted by the number of engineers. The second-stage estimates are aggregate state-level estimates of hours worked calculated from the predicted hours worked in the individual model, which are then aggregated to the state level. The ACS sample uses individuals who earn less than $250 per hour and who are college graduates. Standard errors are in parentheses.

***Significant at the 1 percent level.

**Significant at the 5 percent level.

*Significant at the 10 percent level.

In the empirical section, we show that licensing for these occupations has grown somewhat more rigorous during the period 2001–2012. We then estimate a panel-data model for the engineers in our sample using the ACS. Our estimates show a small influence of occupational licensing on both wages and employment in a variety of specifications and sensitivity analysis tests. In the U.S. economy, if engineers achieve the goal of their professional association of more rigid requirements, and a longer time to become an engineer, the growth of regulation of the occupation may reduce customer access to engineers and slow down the ability of builders and manufacturers to use regulated engineering services. Our study provides a first look at these issues. Exploring the potential issue of selection across engineering specialties, and using more detailed analysis such as the use of discontinuities when the passage of more rigorous laws occurs, may provide more refined or precise estimates and examples of the role of regulation in the market for engineers. Further, a more thorough analysis would include individuals who have attained a license rather than licensing coverage, and these data would allow us to obtain a better measure of the influence of occupational licensing on those who chose to get the credential to legally do certain engineering tasks. Nevertheless, engineers may be seeking greater control over the supply of engineers, but have not attained the level of labor market control that other occupations such as dentists or electricians have been able to gather over time. When the licensing of engineers influences not only the broad coverage of the law in the field, but also the occupational attainment of workers in the profession, they can obtain what many other licensed professionals have been able to acquire in the labor market.

Appendix

Table 7A.1   Influence of professional exams (PE) and pass rates on wage determination of engineers

	All engineers		Civil engineers		Electrical engineers		Industrial engineers	
	(1)	(2)	(3)	(4)	(5)	(6)	(7)	(8)
Analysis Variables	One-step Log of wage	Two-step Log of wage	One-step Log of wage	Two-step Log of wage	One-step Log of wage	Two-step Log of wage	One-step Log of wage	Two-step Log of wage
PE * pass rate	−0.008***	0.000	−0.005***	0.001	−0.013***	0.000	−0.006***	0.000
	(0.000)	(0.001)	(0.000)	(0.002)	(0.000)	(0.002)	(0.000)	(0.001)
PE	0.452***	−0.040	0.251***	−0.044	0.744***	−0.027	0.320***	−0.066
	(0.005)	(0.048)	(0.007)	(0.092)	(0.008)	(0.096)	(0.011)	(0.074)
Pass rate	0.004***	−0.001	0.002***	−0.001	0.008***	−0.000	0.003***	−0.000
	(0.000)	(0.001)	(0.000)	(0.001)	(0.000)	(0.001)	(0.000)	(0.001)
Constant	3.210***	3.345***	3.340***	3.307***	3.051***	3.418***	3.156***	3.295***
	(0.005)	(0.040)	(0.007)	(0.077)	(0.008)	(0.081)	(0.011)	(0.062)
Observations	2,195,841	123	1,007,204	123	775,218	123	413,419	122
R^2	0.009	0.976	0.004	0.918	0.019	0.924	0.004	0.940
Basic control	No	Yes	No	Yes	No	Yes	No	Yes
Year fixed	No	Yes	No	Yes	No	Yes	No	Yes
State fixed	No	Yes	No	Yes	No	Yes	No	Yes

Note: Estimated with age, schooling in years, gender, marital status, experience, experience squared, race, U.S. citizenship, for-profit sector, and self-employment. Two-stage regressions are weighted by the number of engineers. The second-stage estimates are aggregate state-level estimates of hours worked calculated from the predicted hours worked in the individual model, which are then aggregated to the state level. The ACS sample uses individuals who earn less than $250 per hour and who are college graduates. Standard errors are in parentheses.

***Significant at the 1 percent level.

**Significant at the 5 percent level.

*Significant at the 10 percent level.

Table 7A.2   Effect of professional exams (PE) and pass rates on hours worked

Analysis Variables	All engineers		Civil engineers		Electrical engineers		Industrial engineers	
	(1) One-step Hours worked	(2) Two-step Hours worked	(3) One-step Hours worked	(4) Two-step Hours worked	(5) One-step Hours worked	(6) Two-step Hours worked	(7) One-step Hours worked	(8) Two-step Hours worked
PE * pass rate	0.000***	−0.000	0.001***	0.000	−0.000***	−0.000	0.001***	−0.000
	(0.000)	(0.000)	(0.000)	(0.000)	(0.000)	(0.000)	(0.000)	(0.000)
PE	−0.018***	0.016**	−0.061***	0.006	0.041***	0.005	−0.033***	0.023
	(0.002)	(0.007)	(0.002)	(0.010)	(0.002)	(0.015)	(0.003)	(0.014)
Pass rate	0.000***	0.000	−0.000***	0.000	0.001***	0.000*	0.000*	−0.000
	(0.000)	(0.000)	(0.000)	(0.000)	(0.000)	(0.000)	(0.000)	(0.000)
Constant	3.750***	3.751***	3.773***	3.742***	3.709***	3.752***	3.775***	3.779***
	(0.001)	(0.006)	(0.002)	(0.008)	(0.002)	(0.013)	(0.003)	(0.012)
Observations	2,195,841	123	1,007,204	123	775,218	123	413,419	122
R^2	0.002	0.950	0.004	0.952	0.003	0.882	0.003	0.915
Basic control	No	Yes	No	Yes	No	Yes	No	Yes
Year fixed	No	Yes	No	Yes	No	Yes	No	Yes
State fixed	No	Yes	No	Yes	No	Yes	No	Yes

Note: Estimated with age, schooling in years, gender, marital status, experience, experience squared, race, U.S. citizenship, for-profit sector, and self-employment. Two-stage regressions are weighted by the number of engineers. The second-stage estimates are aggregate state-level estimates of hours worked calculated from the predicted hours worked in the individual model, which are then aggregated to the state level. The ACS sample uses individuals who earn less than $250 per hour and who are college graduates. Standard errors are in parentheses.

***Significant at the 1 percent level.

**Significant at the 5 percent level.

*Significant at the 10 percent level.

Table 7A.3   Effect of pass rates on hours worked

Analysis Variables	All engineers		Civil engineers		Electrical engineers		Industrial engineers	
	(1) One-step Hours worked	(2) Two-step Hours worked	(3) One-step Hours worked	(4) Two-step Hours worked	(5) One-step Hours worked	(6) Two-step Hours worked	(7) One-step Hours worked	(8) Two-step Hours worked
PE	−0.055	0.116	0.020	0.224	−0.737***	−0.690***	0.592***	0.496***
	(0.038)	(0.096)	(0.054)	(0.158)	(0.080)	(0.159)	(0.075)	(0.151)
Pass rate	−0.006***	−0.006***	0.013***	0.014***	−0.017***	−0.018***	−0.037***	−0.039***
	(0.001)	(0.002)	(0.001)	(0.003)	(0.001)	(0.003)	(0.002)	(0.003)
Observations	2,601,433	123	1,175,107	123	908,387	123	517,939	122
R^2	0.038	0.962	0.051	0.946	0.045	0.953	0.038	0.967
Basic control	Yes	Yes	Yes	Yes	Yes	Yes	Yes	Yes
Year fixed	Yes	Yes	Yes	Yes	Yes	Yes	Yes	Yes
State fixed	Yes	Yes	Yes	Yes	Yes	Yes	Yes	Yes

Note: Basic controls include age, schooling in years, gender, marital status, experience, experience squared, race, U.S. citizenship, for-profit sector, and self-employment. Two-stage regressions are weighted by the number of engineers. The second-stage estimates are aggregate state-level estimates of hours worked calculated from the predicted hours worked of individuals, which are then aggregated to the state level. Standard errors are in parentheses.

***Significant at the 1 percent level.

**Significant at the 5 percent level.

*Significant at the 10 percent level.

References

Brinegar, P., and K. Schmitt. 1992. "State Occupational and Professional Licensure." In *The Book of the States, 1992–1993*, 567–80. Lexington, KY: Council of State Governments.

Council of State Governments. 1952. *Occupational Licensing Legislation in the States*. Chicago: Council of State Governments.

Dunlop, J. T. 1958. *Industrial Relations Systems*. New York: Holt.

Folk, H. 1970. *The Shortage of Scientists and Engineers*. Lexington, MA: Lexington Press.

Freeman, R. B. 1976. "A Cobweb Model of the Supply and Starting Salary of New Engineers." *Industrial and Labor Relations Review* 29 (2): 236–48.

Friedman, Milton. 1962. *Capitalism and Freedom*. Chicago: University of Chicago Press.

Gittleman, M., and M. M. Kleiner. 2016. "Wage Effects of Unionization and Occupational Licensing Coverage in the United States." *Industrial and Labor Relations Review* 69 (1): 142–72.

Haber, S. 1991. *The Quest for Authority and Honor in the American Professions, 1750–1900*. Chicago: University of Chicago Press.

Hansen, W. L. 1961. "The Shortage of Engineers." *Review of Economics and Statistics* 43 (3): 251–56.

Hirsch, B. T., and D. A. Macpherson. 2011. "Union Membership and Coverage Database from the Current Population Survey." September. http://www.unionstats.com.

Kleiner, M. M. 2006. *Licensing Occupations: Ensuring Quality or Restricting Competition?* Kalamazoo, MI: W. E. Upjohn Institute.

———. 2013. *Stages of Occupational Regulation: Analysis of Case Studies*. Kalamazoo, MI: W. E. Upjohn Institute.

Kleiner, M. M., and A. B. Krueger. 2010. "The Prevalence and Effects of Occupational Licensing." *British Journal of Industrial Relations* 48 (4): 676–87.

———. 2013. "Analyzing the Extent and Influence of Occupational Licensing on the Labor Market." *Journal of Labor Economics* 31 (2): S173–202.

Kleiner, M., and R. Kudrle. 2000. "Does Regulation Affect Economic Outcomes? The Case of Dentistry." *Journal of Law and Economics* 43 (2): 547–82.

Kleiner, M., and R. Todd. 2009. "Mortgage Broker Regulations That Matter: Analyzing Earnings, Employment, and Outcomes for Consumers." In *Studies of Labor Market Intermediation*, edited by D. Autor. Chicago: University of Chicago Press.

Lewis, H. Gregg. 1986. *Union Relative Wage Effects: A Survey*. Chicago: University of Chicago Press.

Perloff, J. M. 1980. "The Impact of Licensing Laws on Wage Changes in the Construction Industry." *Journal of Law and Economics* 23 (2): 409–28.

Pfatteicher, S. K. A. 1996. "Death by Design: Ethics, Responsibility and Failure in the American Civil Engineering Community, 1852–1986." PhD diss., University of Wisconsin–Madison.

Princeton Data Improvement Initiative (PDII). 2008. http://www.krueger.princeton.edu/PDIIMAIN2.htm.

Rubin, D. K., and J. L. Tuchman. 2012. "Professional Engineers' License Debate Grows in Intensity." *Engineering News-Record* Oct. 17, pp. 1–2.

Russell, J., and W. B. Stouffer. 2003. "Change Takes Time: The History of Licensure and Continuing Professional Competency." Report, American Society of Civil Engineers http://www.enr.com/articles/2911-professional-engineers-license-debate-grows-in-intensity?v=preview.

U.S. Department of Labor, Bureau of Labor Statistics. 2013. Current Population Survey.

8

Dynamics of Engineering Labor Markets
Petroleum Engineering Demand and Responsive Supply

Leonard Lynn, Hal Salzman, and Daniel Kuehn

8.1 Introduction

The dynamics of engineering labor markets are controversial. Some believe they are similar to other labor markets, in which the supply of workers is responsive to demand as reflected in salaries. As the demand for engineers increases, salaries increase, motivating more students to major in engineering and more incumbents to stay in engineering. Others assert that the specialized knowledge and arduous education and training required of engineers inherently limit the size of the labor pool. According to this view, the U.S. education system does not produce a sufficient number of qualified engineers to meet national needs. One problem is said to be a lack of interest in the profession of engineering by American young people, especially women and some minorities. Another is a supposed weakness in the U.S. K–12 education system which, it is claimed, does not produce a sufficient number of high school graduates qualified to enter university engineering programs.

Leonard Lynn is Professor Emeritus at Case Western Reserve University. Hal Salzman is Professor of Planning and Public Policy at the Edward J. Bloustein School of Planning & Public Policy and Senior Faculty Fellow at the John J. Heldrich Center for Workforce Development at Rutgers University. Daniel Kuehn is research associate at The Urban Institute.

The research was supported through grants from the Alfred P. Sloan Foundation and the National Science Foundation (Human and Social Dynamics Program, no. SES-0527584; Social Dimensions of Engineering, Science, and Technology no. 0431755). Research assistance was provided by Purba Ruda. Parts of this chapter are drawn from Lynn and Salzman (2010) and Salzman and Lynn (2010). Professor Thomas K. Holley, Director of the Petroleum Engineering Program at the University of Houston, was also kind enough to read a draft of this chapter and make suggestions and comments. The authors, of course, take full responsibility for the chapter's conclusions and any factual errors that might remain in it. For acknowledgments, sources of research support, and disclosure of the authors' material financial relationships, if any, please see http://www.nber.org/chapters/c12693.ack.

Petroleum engineering provides an illustrative case study for assessing the dynamics of engineering labor markets more generally. Petroleum engineering is a field where concerns about shortages of talent have been strongly voiced by industry leaders for a number of years, first prompted by observations over the past decade of labor force demographics indicating a need for increased hiring. The anticipated wave of retirements in the industry, in a workforce where half of all geophysicists and engineers are expected to retire by 2018, is characterized by people in the oil and gas industry as "the great crew change." The demand for replacement hiring was further compounded by increased exploration that requires hiring additional petroleum engineers. The history of petroleum engineering demand reflects the cyclical demand for engineers in response to changes in the industry.

The building of the Trans-Alaska Pipeline and increased oil exploration in other regions led to rapidly increasing demand for petroleum engineers in the 1970s. But once the pipeline was built and new domestic oil exploration slowed, demand for new petroleum engineering graduates fell off. However, by the late 1990s many of the Trans-Alaska Pipeline generation engineers were approaching retirement age and, in the early 2010s, more of the workforce passed retirement age just as spikes in oil prices were leading to an anticipated increase in need for petroleum engineers as firms were motivated to explore more aggressively for new oil fields.

In this chapter, we first examine the recent history of claims that the market for engineers is dysfunctional. We then investigate the adjustment of the supply of engineering graduates to meet sharp increases in demand for petroleum engineers, and the implications for other areas of engineering and for the supply and demand of science and engineering (S&E) personnel. This investigation begins with a description of the field of petroleum engineering. We describe the field of petroleum engineering and supply responses to the market signals of the late 1990s and early in the first decade of the twenty-first century. We then discuss the mechanisms that signaled an upcoming demand spike to universities and other institutions, and the responses that allowed those institutions to meet the demand. We conclude with the implications of this case for the shortage and mismatch claims made about the S&E workforce.

8.2 Shortage and Mismatch Claims

Following the Great Recession of 2008, a group of technology company CEOs and others on the President's Council on Jobs and Competitiveness called for special measures by government, business, and universities to increase the number of U.S. engineering graduates by 10,000 a year (President's Council on Jobs and Competitiveness 2011). The Institute for Electrical and Electronics Engineering President Ron Jensen supported this call, asserting that more engineering graduates are needed because engineers drive innovation and create jobs (IEEE 2011). Companies cited a shortage of engineers as the reason they have "had" to move work off-shore. CBS News featured Andrew Liveris, president of Dow Chemical, lamenting the scarcity of qualified engineers in the United States, saying this had caused his company to open research and development labs in Brazil, China, India, and Eastern Europe instead of the United States (CBS News 2011). And famously in 2011, Steve Jobs told President Obama that the reason he located 700,000 manufacturing jobs in China instead of the United States was his inability to find enough industrial engineers in the United States (Isaacson 2011; Salzman 2013, 59). These calls for government to increase the number of engineering graduates follow a decade-long series of reports and policy statements decrying shortages of science and engineering graduates.[1]

Some claim that persistent high rates of unemployment in the wake of the economic crisis that started in 2008 were caused by a mismatch between the skills of the unemployed and the skill needs of the new economy. Rapidly advancing technology implies that the skill needs of this "new" economy are primarily in the area of S&E human resources. These claims are compounded by the fear that the United States is losing (or will soon be losing) a technological race with other countries. This fear is supported by statistics showing that China and India produce hundreds of thousands of engineers.[2] Furthermore, numerous reports argue that the United States is losing technological competitiveness because of weaknesses in our K–12 math and science education system (see Salzman and Lowell [2008] and Lowell and Salzman [2007] for critique). One alleged consequence of this weakness is a shortage of Americans well enough educated to succeed in university engineering programs.[3]

These concerns have motivated proposals for heavy investments to remedy the weaknesses in the United States' K–12 math and science education system, make engineering as a career seem more attractive to young people (especially women and minorities who are underrepresented in engineering), offer more scholarships to engineering students, and expand university engineering programs. These primary and secondary educational reforms are typically complemented by proposals to bring in larger numbers of talented foreign engineers while retaining more of those who are already here at our universities and in our high-tech workforce.

1. For a history of STEM shortage claims, see Teitelbaum (2014) and Salzman (2013).
2. Although China and India, each with a population of over a billion people, do graduate many more engineers than the United States, the cited numbers have been found to overstate the actual numbers of bachelor's degree level engineers with globally competitive skills and, more importantly, reflect the domestic needs of those countries for their vastly greater demand than in the United States for infrastructure engineering; engineering infrastructure represents the largest share of demand for engineers in nearly all countries (Lynn and Salzman 2010).
3. The most prominent publications making these claims are probably those of the National Research Council (2007, 2010).

The purported need to increase the number of engineering graduates available in the United States for employers is based on a number of assumptions. It is assumed that there is indeed a shortage of engineers that cannot be met by the normal functioning of labor markets. Furthermore, it is assumed that the size of the stock of engineers in a country is proportional to the country's economic and military security, or even that to be secure a country must have more engineers than its rivals. Still another assumption is that increasing the supply of engineers, regardless of the demand expressed in the marketplace, will increase innovation and in turn drive economic growth. Still more assumptions underlie the proposals to fix the supposed shortage problem, such as the notion that part of the claimed shortage of engineering graduates is caused by the failure of American schools to train K–12 students so that they are qualified to enter university engineering departments, and this failure is exacerbated by a failure to convince students about the excitement of engineering as a profession. The contention, then, is that there are "market failures" leading to shortages in the supply of engineers in the United States. U.S. universities and students, according to this logic, are not responding to the national need for larger numbers of well-educated engineers, which requires the country to bring in larger numbers of talented foreign professionals and retain those who are here at our universities (Lynn and Salzman 2010).

Many labor market analysts have been skeptical of these claims, arguing that much of the push to train (or import) more engineers is actually motivated by the interests of employers in lowering labor costs rather than actual labor market dysfunctions. If there is a shortage of engineering graduates, companies can pay new graduates more, attracting additional engineers in the future and using market wages to allocate the existing workers to firms in greatest need. If there is no shortage, the costs to society of creating an oversupply of engineers are high. These include the wasted time and efforts of bright young people being trained for jobs that do not exist and the wasted resources of government and universities creating the capacity to train more engineers than are needed.[4] Moreover an extreme oversupply might dampen the interest of new generations of Americans in pursuing high-tech education, leading to real shortages in the future (Teitelbaum 2014, 118–54).

How responsive, generally, are labor markets to rapid changes in demand? Several studies have considered the labor market response to a large demand shock following the discovery of natural resources. This work typically finds evidence that labor markets are very responsive, although the response lags sharp initial wage increases. Indeed, recent natural resource booms such

---

4. Some of the reports and papers arguing for action to increase the number of engineering graduates in the United States are the Council on Competitiveness (2005), Farrell and Grant (2005), National Research Council (2007, 2012), and the National Association of Manufacturers (2005). We have addressed these arguments in Lowell and Salzman (2007), Lynn and Salzman (2010), Salzman (2013), and elsewhere.

as the building of the Trans-Alaska Pipeline (Carrington 1996) have been associated with higher labor supply elasticity than older cases such as the California Gold Rush (Margo 2000). There is also a literature that provides critical insights into the general equilibrium response to a natural resource shock (e.g., Marchand 2012; Clay and Jones 2008; Aragon and Rud 2013; Black, McKinnish, and Sanders 2005) that considers the broader labor market rather than engineering per se.

So far, evaluations of arguments about alleged market failures in engineering job markets have generally relied on research showing the numbers of graduates each year who are hired into engineering jobs, university enrollments, salary trends, and other statistical indicators for engineers in the aggregate. Dynamic changes in a specific labor market are more difficult to study, yet such studies are needed to give us a better sense of what influences the supply of engineers (Meiksins and Smith 1996). How quickly does the education system, for example, respond to changes in market demand? Do shortages of qualified engineering students or institutional rigidities in universities cause failures in supply to meet the demand for engineers? Let us now turn to the case of the market response to sudden changes in the demand for petroleum engineers.

8.2.1   Petroleum Engineering: What Do Petroleum Engineers Do?

Petroleum engineers are engaged in a wide range of activities related to the development and exploitation of crude oil and natural gas fields. Their activities span the life cycle of the fields: finding reservoirs, deciding how to get the best yield from them, designing equipment for drilling and pumping, ensuring regulatory compliance, getting additional yield from older fields, and shutting down depleted fields. Major areas of specialization within petroleum engineering include drilling engineering, production engineering, reservoir engineering, and petrophysical engineering (designing tools and techniques to determine rock and fluid characteristics). Engineers from mechanical, civil, electrical, geological, and chemical engineering have contributed to these fields. The *Princeton Review* (2013) says the work "can mean travel, long stays in unusual (and sometimes inhospitable) locations, and uncertain working conditions." The Review cites a petroleum engineer as saying, "If you're into engineering and gambling, petroleum engineering is for you." Because of the sometimes harsh working conditions and sometimes quickly fluctuating employment, prospects' starting salaries have tended to be high compared to other fields of engineering. While a bachelor's degree in petroleum engineering is preferable for petroleum engineers, some hold degrees in mechanical or chemical engineering (Bureau of Labor Statistics 2014).

The profession of petroleum engineering got its formal start early in the twentieth century when it became clear that the harvesting of surface oil and the use of water well-drilling techniques were not sufficient to meet the burgeoning demand for oil and gasoline. The first oil fields were in Pennsylvania, and the University of Pittsburgh introduced courses in oil and gas industry practices in 1910. The first degree in petroleum engineering was granted in 1915. The University of California at Berkeley introduced courses in petroleum engineering around the same time, and established a four-year petroleum engineering program in 1915. Around twenty universities now offer petroleum engineering education programs; the largest programs (in number of bachelor of science degrees awarded in recent years) are Texas A&M University, Pennsylvania State University, Colorado School of Mines, University of Texas at Austin, and Texas Tech University. A number of programs at other universities have not awarded degrees in recent years, and some programs have capped enrollments. Some companies offer in-house training programs in petrochemical engineering.

At first the focus of petroleum engineers was on finding ways to address drilling problems. In the 1920s it shifted to improving well design and production methods. A decade later, a major concern was finding ways to maximize outputs from entire fields. After World War II, petroleum engineers improved the techniques of reservoir analysis and petrophysics (American Petroleum Institute 1961). New technology also was needed to support the new offshore oil industry. More recently, additional challenges have been posed by the desire to find oil and natural gas in the arctic, very deep water, and desert conditions. These have required additional technical inputs from thermohydraulics, geomechanics, and intelligent systems. Still another development has been the increased use of hydraulic fracking techniques for the extraction of hydrocarbons.

This chapter draws on two data sources to assess the size of the petroleum engineering workforce: the American Community Survey (ACS), produced by the Census Bureau (Ruggles et al. 2017), and the Occupational Employment Statistics (OES), produced by the Bureau of Labor Statistics (1996, 1998, 2004, 2006, 2010, 2014).[5] In 2013 there were about 35,000 petroleum engineers in the United States (37,340 according to the ACS and 34,910 according to the OES; see figure 8.2). The ACS estimates show a 58 percent increase from the 23,604 petroleum engineers in the workforce in 2003. Employers include major oil companies and oil-industry suppliers such as Exxon-Mobil, Chevron, Haliburton, and Schlumberger, as well as large numbers of smaller independent oil, services, and production companies. In 2015 the mean salary for petroleum engineers in the United States was estimated to be $135,985 (in 2016 USD) according to the ACS and $151,477 according to the OES data sets. Median salaries were somewhat lower in both data sets at $106,325 and $131,630, respectively.

In the number of bachelor's degrees awarded, petroleum engineering is

5. The ACS is a population survey (workers), while the OES is based on employer surveys, which may result in diverging estimates due to sampling and occupational/industry definitional differences, as well as response differences between workers and employers (Abraham et al. 2013).

one of the smaller engineering fields. In 2014–2015, only 1,688 bachelor's degrees of 87,812 bachelor's degree awards in engineering in the United States were petroleum engineers. Nearly half the bachelor's degrees awarded in engineering went to students in just three fields: mechanical (25 percent), civil (18 percent), and electrical (21 percent). In 2012–2013, some 14 percent of new petroleum engineers were women, slightly less than the 19 percent of women in engineering overall. The number of bachelor's degrees awarded in petroleum engineering more than quintupled from 2003 to 2015, from 252 to 1,383. Petroleum engineering has the largest rate of increase of any engineering field, though the even smaller workforces in mining engineering and nuclear engineering also had large rates of increase.[6]

8.3   Market Signals and Market Responses

In the 1970s, the building of the Trans-Alaska Pipeline and increased oil exploration in other regions led to rapidly increasing demand for petroleum engineers. By 2002 however, *Occupational Outlook* forecast an employment decline "because most of the petroleum-producing areas in the United States already have been explored" (Bureau of Labor Statistics 2004), and this continued to be the forecast through the 2008 edition of *Occupational Outlook*. In the 2014–2015 edition, however, the Bureau of Labor Statistics forecast for 2012–2022 changed to a projected employment increase of 26 percent over the coming decade because "petroleum engineers increasingly will be needed to develop new resources, as well as new methods of extracting more from existing sources." The shift to greater exploration followed the 2008 oil price spike, which also increased the returns to investments in types of oil extraction that were previously cost prohibitive (e.g., horizontally drilled and hydraulically fractured shale), thus increasing the demand for petroleum engineers, especially those with new skill sets. Following this period of rapid expansion, oil prices and employment fell, and growth moderated. The 2014–2024 projections, in the 2016–2017 edition of the *Occupational Outlook*, were substantially lower, forecasting a more modest 10 percent growth over the period, with the petroleum engineering employment having fallen from 38,500 in 2012 to 35,100 in 2014.

The number of job openings began to exceed the number of graduates around 2002, even though there still had been no overall workforce growth.[7] This was because of retirements and because there had been little hiring since the earlier oil boom and hiring expansion of the 1970s and 1980s. In

6. Although no other field quadrupled, mining engineering went from 85 to 231, and nuclear engineering went from 202 to 614 from 2004 to 2013 (Yoder 2013).

7. The BLS's *Occupation Outlook Handbook* in 2002 noted that "Employment of petroleum engineers is expected to decline through 2010 because most of the potential petroleum-producing areas in the United States already have been explored. Even so, favorable opportunities are expected for petroleum engineers because the number of job openings is likely to exceed the relatively small number of graduates. All job openings should result from the need to replace petroleum engineers who transfer to other occupations or leave the labor force."

interviews with managers in oil companies, we found high levels of concern because the large cohort of engineers hired in the 1970s and 1980s was retiring just as the firms were launching large development and maintenance projects. This underlying demand was then exacerbated by the oil price spike, which intensified exploration efforts, as higher oil prices made previously unprofitable exploration (which in many cases posed greater engineering challenges) now profitable.

The response to this confluence of events—little hiring for many years, a current workforce that was aging and retiring, and a sudden increase in oil exploration—led to an observable demand for new petroleum engineers that exceeded the number of graduates each year. The earlier demand pressure from retirements had already led to increases in starting salaries, but with the oil-price spike petroleum engineering starting salaries rose even further, becoming the highest of all fields of engineering for new bachelor's degree graduates (National Association of Colleges and Employers 2010). Starting salaries (in 2014 dollars) jumped from an already high $65,024 in 1997 to $72,485 in 1999 and then rose to $73,029 and $75,598 in 2003 and 2005, respectively, only to fall to $73,711 in 2008 (Bureau of Labor Statistics 2004, 2006; National Association of Colleges and Employers 2009). The 2010s increase in demand for petroleum engineering graduates began in 2009, when real starting salary offers jumped to $91,275 and steadily increased afterward, reaching a peak of $99,111 in 2013 (National Association of Colleges and Employers 2010, 2014). In all these years, petroleum engineering salaries were higher than other engineering salaries but, until the spike in demand, the petroleum engineering starting salary premium was relatively small. For example, the 1997 $43,674 starting salary for petroleum engineers was only slightly greater than that for the second-highest-paid engineering field, chemical engineers, who received an average starting salary of $42,817. In 2010, however, the nominal starting salary of $86,220 for petroleum engineers was much higher than that of the second-highest field, still chemical engineering, which was only $65,142 (National Association of Colleges and Employers 2010). Petroleum engineer starting salary figures from the National Association of Colleges and Employers (NACE) are reported in figure 8.1, along with mean salary trends for all petroleum engineers estimated using the ACS and the OES database (all in 2016 dollars). Starting salary data from NACE should be interpreted with some caution as they are not comprehensive, but are based on a survey of only NACE-member employers and a small sample of petroleum engineers. However, the trends in starting salaries track trends in the salaries of all petroleum engineers, increasing impressively over the last decade. From 2003 to 2013, petroleum engineer starting salaries reported by NACE and mean petroleum engineer salaries reported in the OES both grew by over 35 percent. The ACS reports somewhat slower mean salary growth of just under 25 percent.

Employment growth between 2003 and 2013 was even more substantial

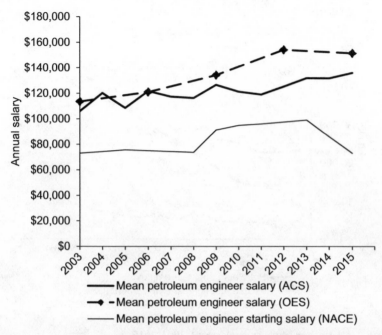

Fig. 8.1   Mean real petroleum engineer salaries and starting salaries, 2003–2015

Source: Authors' calculations from annual NACE salary surveys, BLS Occupational Employment Statistics, and the American Community Survey. Starting salaries for 2004, 2006, 2007, and 2012 were not available and have been linearly interpolated. The OES data are three-year moving averages; only one data point in each three-year series is plotted.

than earnings growth. Figure 8.2, which presents data on petroleum engineer employment and the number of foreign petroleum engineers working in the United States and petroleum engineer employment from the OES, shows that employment grew even more rapidly between 2003 and 2013 than earnings. The ACS reports a 58 percent increase in petroleum engineer employment over this decade while the OES reports a 200 percent increase. The major reason for the difference in percentage change is a large difference in the estimated 2003 employment due, we believe, to the relatively small samples used to estimate national employment levels, and the OES uses a three-year moving average, which is a lagged estimate of annual growth.[8] Nevertheless, the data sources consistently show substantial growth from a workforce of around 20,000 to one of around 35,000 (recent surveys from ACS and OES show similar estimates of workforce sizes in 2013 and then diverge dramatically after 2013, likely due to ACS survey differences; the

8. In the ACS, only several hundred survey respondents report working as petroleum engineers each year.

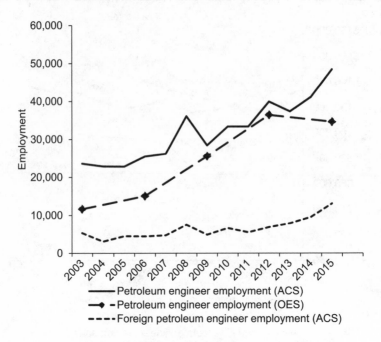

Fig. 8.2   Petroleum engineer employment, 2003–2015
Source: Authors' calculations from the BLS Occupational Employment Statistics and the American Community Survey. The OES data are three-year moving averages; only one data point in each three-year series is plotted.

large differences in the two surveys mentioned above are for earlier years).[9] The increase in the number of foreign (i.e., noncitizen) petroleum engineers during this period in the ACS was somewhat slower than the growth in number of all petroleum engineers (48 percent growth compared to 58 percent).

Rapid growth in the employment of petroleum engineers generally came by hiring younger workers. Figure 8.3 shows the change in the age distribution of the workforce in 2003 and the age distribution in 2013. While the

9. The OES is an establishment survey using annual estimates that are a moving average of the previous three years, whereas the ACS is a household survey. The sharp divergence in employment after 2013 may be the result of survey and methodological differences between the two surveys but, we postulate, also could be an artifact of the decline in regular industry employment. As an establishment survey, the OES reports wage employment of petroleum engineers engaged in engineering activities by firms; increases in retirements, layoffs, and contract hiring would lead to lower OES but not lower ACS employment reporting if those leaving firms become consultants or contract workers, and particularly if they are hired as management consultants rather than performing specific engineering tasks (personal communication with OES staff, June 20, 2017). Moreover, to the extent that incumbent and older workers retire or are laid off at a greater rate than employment of younger workers, and become consultants or contract workers, and/or younger workers are hired as contract workers or through staffing firms, this would lead to an increase in overall employment of "petroleum engineers" in population surveys but a decline in firm-based employment numbers. The OES data are consistent with the numerous reports from universities and firms about employment patterns (SPE 2015; Weaver 2017; Gallucci 2016).

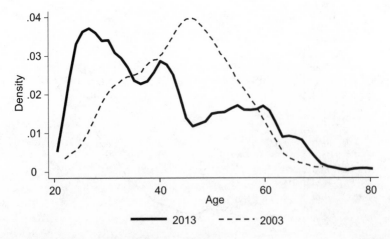

kernel = epanechnikov, bandwidth = 1.4544

Fig. 8.3   Petroleum engineer age distribution, 2003 and 2013
Source: Authors' calculations from the American Community Survey.

modal age of petroleum engineers in 2003 was just under fifty, by 2013 it was under thirty. Although some of these new employees could have been pulled in from other sectors of the economy, many are new graduates of petroleum engineering programs and other engineering programs such as chemical engineering, which exhibited considerable growth over this period.[10]

The changes in earnings and employment between 2003 and 2013 suggest *a tremendous responsiveness of employment to wages*, with an implicit elasticity of 2.4 from the ACS and 5.7 from the OES data.[11] Perhaps because petroleum engineering is a relatively small occupational group, it shows a response of this magnitude to price signals, as even a modestly sized absolute change has a large proportionate effect.

In response to increasing median salaries, starting salaries, and other market signals, the number of new petroleum engineering bachelor's degrees awarded by U.S. universities more than quintupled between 2003 and 2015. Most of the growth was concentrated between 2008 and 2011, at the same time that the strongest starting-salary growth was occurring. During this period, petroleum engineering bachelor's awards grew from 521 to 1,030.

---

10. A degree in mechanical or chemical engineering may also suffice for employment as a petroleum engineer according to the *Occupational Outlook Handbook* (Bureau of Labor Statistics 2014).

11. While these figures strongly indicate a responsive labor market, the exact point estimates should be interpreted with considerable caution. Changes in employment and earnings over the course of ten years suggest that these are relatively long-run elasticities, which are expected to be higher. They also pertain to a small workforce, for which rapid growth is easier to accommodate. Finally, the elasticity estimates are clearly imprecise and vary considerably across data sources.

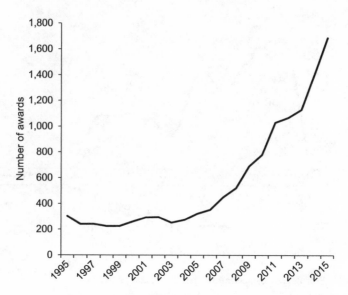

Fig. 8.4   Bachelor's degrees awarded in petroleum engineering, 1995–2015
Source: Authors' calculations from the IPEDS.

Texas A&M and Colorado School of Mines more than tripled their output of new graduates from 42 to 128 and 32 to 100, respectively. As shown in figure 8.4, the dramatic increase in petroleum engineering bachelor's degrees awarded followed the rise in starting salaries, which in turn reflected an increase in industry demand. This would seem to be a clear textbook case of efficient and responsive market functioning. It seems to show that normal market mechanisms, namely wage increases, can dramatically and quickly increase supply.

8.4   Dependencies on Domestic- or Foreign-Student Supply

A key claim about the U.S. S&E workforce is that it is dependent on foreign students and workers because it is not possible to find sufficient numbers of U.S. S&E workers. However, when we examine the dramatic increase in petroleum engineering graduates we find, interestingly, that although a significant source of supply for petroleum engineers historically has been foreign students, the new demand accompanied by sharp increases in salaries resulted in markedly increased numbers of domestic students (U.S. citizens and permanent residents) responding to these market signals and graduating in petroleum engineering. In the initial stages of increased hiring and large salary increases, in the middle of the first decade of the twenty-first century, nearly the entire increase in graduates was composed of U.S. students (citizens and permanent residents), and the share of foreign students declined.

Toward the end of the decade, the number and share of foreign students increased, but this was also a period when salary growth began to slow and even plateau. The share of petroleum engineers who were foreign born ranged between 16 percent and 21 percent during the period of strong wage and employment growth, and then increased sharply after 2012 as annual wage and employment growth weakened or declined. As will be discussed below, some petroleum engineering department chairs are now expressing concerns about an impending oversupply of new graduates.

In terms of understanding responsiveness of engineering labor markets, it is important to note that it is not just the overall supply of petroleum engineering graduates from colleges that appears to have been responsive to demand and wages, but it is the *domestic supply in particular* that supplies the increased pool of graduates. As wages increased and job demand in the United States increased, there has been a shift in the relative share of domestic students in the graduating pool. The percentage of foreign petroleum engineering graduates in the United States on student visas, the highest of any of the engineering fields at the bachelor's degree level, declined from a peak of 34 percent in 2005 to 24 percent in 2013 as the domestic supply increases (figures 8.5 and 8.6). At the bachelor's degree level, the

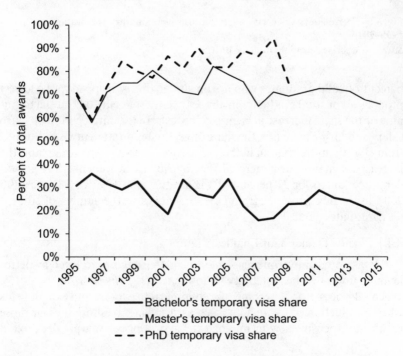

Fig. 8.5   Temporary visa share of degrees awarded in petroleum engineering, 1995–2015

Source: Authors' calculations from the IPEDS.

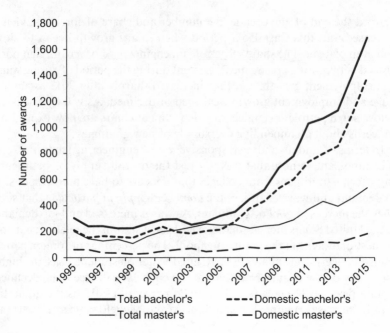

Fig. 8.6 Petroleum engineering degree awards by temporary visa status, 1995–2015

Source: Authors' calculations from the IPEDS.

percent of total graduates who are on student visas dropped to the lowest proportion of total graduates in almost twenty years in 2007, at the beginning of the rapid increase in bachelor's awards. In the initial period of sharp salary and hiring increases, the share of graduates on student visas dropped from slightly more than half the proportion twelve years earlier, and the increased demand was largely satisfied by American students (foreign students accounted for 31 percent of the graduates in 1995 vs. 17 percent in 2007, as the increase came from U.S. students while the number of student visa graduates held steady).

8.4.1   When Demand for Engineers Drops

As was noted above, one reason for the higher salaries received by petroleum engineers is the instability of job markets (*Princeton Review* 2013). When oil prices drop it becomes uneconomical to explore and exploit fields where production costs are high (such as deepwater offshore sites, or those requiring expensive new technologies).[12] As oil prices sharply declined in

---

12. The relationship is not entirely linear, because petroleum projects differ in the time required for completion. Onshore projects such as exploration through first production can be relatively quick, while deepwater offshore projects can take years before first production.

2014 and early 2015 there were increasing reports of job cuts, either those that had already occurred or those that were feared. In February 2015, Reuters stated that more than 100,000 layoffs worldwide had been reported in the oil industry. Halliburton had announced cuts of 8 percent of its global workforce and Schlumberger was planning to eliminate 7 percent of its workforce (Kemp 2015). The Society of Petroleum Engineers' survey found only two-thirds "of 2015 U.S. petroleum engineering bachelor's degree graduates have found jobs in the oil and gas industry, compared with 95 percent in 2014 and 97 percent in 2012" (SPE 2015). The *Occupational Handbooks* show the petroleum engineering workforce fell from 38,500 in 2012 to 35,100 in 2014.

Bloomberg (Shauk 2015) reported concerns of new petroleum engineering graduates about their job prospects commenting, "Six months ago, a degree in petroleum engineering was a ticket to a job with a six-figure salary. Now it's looking like a path to the unemployment office." The director of undergraduate advising for Texas A&M's Petroleum Engineering Department indicated that students were expressing "definite concern" about the job market.

In March 2015, we sent survey questions to the chairs of thirteen leading U.S. petroleum engineering departments asking about their experiences in adjusting capacity to meet industry demands for new petroleum engineering graduates. Three of the four chairs who responded expressed concerns that U.S. universities had overbuilt their capacity and needed to take stronger actions to control growth so as to avoid a glut of new graduates. The fourth chair was more sanguine, expressing confidence that the demand for petroleum engineers would continue growing so that the increased capacity would be needed.

Despite the recent downturn, industry spokesmen have also continued to argue that large numbers of petroleum engineers will be needed in the next few years. Not discussed, however, was how programs and students should respond in the short term if there are not immediate employment opportunities for current graduates.

8.5   Implications and Conclusions

The case presented here suggests that American universities and American students were highly responsive to market signals when it came to addressing the need for new graduates in petroleum engineering. But was the market responsive "enough"? Conceivably, even bigger increases than the doubling and tripling that occurred would have been desirable. Or perhaps quality standards were dropped in an effort to meet the increased demand. While a systematic analysis of these issues goes beyond the scope of this chapter, the following "Industry Alert" from the Society of Petroleum Engineers (SPE) in 2010 is suggestive.

Environmental and remediation companies of all sizes have a real opportunity to take steps in 2010 to address that shortage of engineering talent expected in the next decade, especially in the United States and Europe.

Key factors that are creating this opportunity:

*An increase in the number of graduates in petroleum engineering programs is creating the largest pool in 20 years of young engineers seeking entry into the oil and gas industry.

*The global recession has caused experienced professionals to postpone retirement, offering a window of opportunity to transfer their knowledge to these new entrants.

*These students can contribute very quickly, and companies that act now can begin developing new entrants into autonomous professionals with the complex decision-making and ability required to exploit advanced technology.

*Scaling back on new graduate recruiting in 2010 could lead to a permanent loss of this talent from the industry, and chill the interest of future engineering students in pursuing careers in the oil and gas industry. (Rubin 2010)

This strongly suggests that even in a peak demand year there was no serious shortage in the availability of new petroleum engineering graduates and that, indeed, there was some concern of future generations being turned away from the field if companies did not proactively hire more new graduates than were immediately needed. The petroleum engineering department chairs responding to our survey seem to confirm this impression.

The potential downside of large increases in the supply of engineers is suggested in a guest editorial written by the current department chair and a former department chair of the Petroleum Engineering Department at Texas A&M University. These authors note:

Between fall 2011 and fall 2012, the number of freshmen in petroleum engineering programs in the United States grew from 1,388 to 2,153, a 55 percent increase in one year. The enrollment pressure we are experiencing at Texas A&M suggests that there will be another increase in freshman enrollment in 2013. We are rapidly heading toward having more than 2,000 bachelor of science degree petroleum engineering graduates per year in the United States. So far, essentially all of our graduates have been receiving job offers, but there is concern that the job market may not grow as fast as enrollment and graduation rates. (Hill and Holditch 2013)

These authors note similarities with an earlier ramping up of the number of petroleum engineering students in the 1980s (though they say that if account is taken of the number of students transferring into petroleum engineering in their sophomore and junior years, the increases are even larger this time). They fear a potential collapse of the job market and suggest that more universities should manage "the unbounded growth in enrollment that is currently occurring." They note that the two departments that have historically been the largest in the United States, Texas A&M University and University of Texas at Austin, have indeed controlled their growth, but complain that other departments have not, and have passed the two Texas schools in size of enrollments. Mississippi State University, for example, reinstated its petroleum engineering program in 2014, having suspended it in 1995, and its Fall 2015 enrollment of sixty-seven far exceeded expected enrollment of twenty-five, just as the job market began its decline (Weaver 2017; Lassetter 2014).

So might there still somehow be an impending crisis demanding special measures to increase the number of graduating engineers in petroleum engineering? Employers continue to voice alarm, but some suggest that in the past such warnings were overstated. Rigzone, a website that posts job notices in the petroleum industry, for example, said in a 2011 article that the "great crew change" is indeed a problem, but comments, "Much like the old story about the boy who cried 'wolf' so many times that nobody would listen when the wolf finally was at the door, statistics confirm that the post–World War II 'baby boom' generation is at the retirement door" (Saunders 2011).

What then, does the case of petroleum engineers suggest about other fields of engineering? First, in petroleum engineering there seemed to be no serious difficulty in getting qualified students once students had the strong incentive of high salaries to enter this field of engineering. Was this because qualified students were drawn from other fields leaving them short of qualified students? While it is difficult to totally dismiss this possibility, we have seen no sign of it in the literature or in discussions with people in the field. The four chairs of leading petroleum engineering departments who responded to an email survey sent out in March 2015 indicated that they have experienced no difficulty in recruiting students. While the department chairs complained of having some difficulties in recruiting as many qualified faculty as they might have liked (this need was met by hiring more practitioners with industry experience), they were able to meet the increases in demand for qualified students, with no lowering of standards. U.S. universities seem to have been remarkably flexible. Third, and perhaps most important, it is not clear that nonmarket signals such as projections of demand by experts (U.S. Department of Labor) or industry spokespeople (who may, after all, have a vested interest in talking up prospects for demand), have done any better in their predictions than the market.

In the case of petroleum engineering we saw no signs of problems caused by weaknesses in the U.S. K–12 education system or the motivation of young people to undertake careers in engineering. It seemed, as well, that the United States was reasonably able to meet its needs for S&E workers through domestic student supply. It is important to note that this analysis of the responsiveness of student supply to market signals does not address issues such as diversity and the underrepresentation of groups such as women and some minorities. Research suggests that, in the area of diversity, the market is not effective and thus a need may exist for programs that increase the interest of minorities and women in some of the STEM careers, and their access to some fields of STEM education (women are the majority of life science majors and have been near parity in the mathematics bachelor's degrees for the past forty years). Nor do we advocate the exclusion of talented foreign STEM workers from the U.S. economy. The findings of this analysis, however, suggest that when it comes to providing an appropriate number of engineers in the United States, the U.S. education system and job market have been highly responsive economic forces, not the failures that alarmists have habitually portrayed.

References

Abraham, K. G., J. Haltiwanger, K. Sandusky, and J. R. Spletzer. 2013. "Exploring Differences in Employment between Household and Establishment Data." *Journal of Labor Economics* 31 (5): S129–72.

American Petroleum Institute. 1961. *History of Petroleum Engineering*. New York: American Petroleum Institute.

Aragon, Fernando, and Juan Pablo Rud. 2013. "Natural Resources and Local Communities: Evidence from a Peruvian Gold Mine." *American Economic Journal: Economic Policy* 5 (2): 1–25.

Black, Dan, Terra McKinnish, and Seth Sanders. 2005. "The Economic Impact of the Coal Boom and Bust." *Economic Journal* 115 (503): 449–76.

Bureau of Labor Statistics. 1996. *Occupational Outlook Handbook, 1996–97, UM-St. Louis Libraries ed.*, derived and modified by Raleigh Muns July 20, 1996. http://stats.bls.gov/ocohome.htm.

———. 1998. *Occupational Outlook Handbook, 1998–99, UM-St. Louis Libraries ed.*, derived and modified by Raleigh Muns Apr. 14, 1998. http://stats.bls.gov/ocohome.htm.

———. 2004. *Occupational Outlook Handbook, 2004–05 ed.* Accessed July 15, 2017. http://www.bls.gov/oco/ocos027.htm.

———. 2006. *Occupational Outlook Handbook, 2006–07 ed.* Accessed July 15, 2017. http://www.bls.gov/oco/ocos027.htm.

———. 2010. *Occupational Outlook Handbook, 2010–11 ed.: Engineers.* Accessed July 15, 2017. http://www.bls.gov/oco/ocos027.htm.

———. 2014. *Occupational Outlook Handbook, 2014–2015 ed.* Accessed July 15, 2017. http://www.bls.gov/oco/ocos027.htm.

Carrington, William. 1996. "The Alaskan Labor Market during the Pipeline Era." *Journal of Political Economy* 104:186–218.

CBS News. 2011. "U.S. Not Producing Enough Engineers." June 14. https://www.cbsnews.com/videos/us-not-producing-enough-engineers/.

Clay, Karen, and Randall Jones. 2008. "Migrating to Riches? Evidence from the California Gold Rush." *Journal of Economic History* 68 (4): 997–1027.

Council on Competitiveness. 2005. *Innovate America: Thriving in a World of Challenge and Change*. Washington, D.C.: Council on Competitiveness.

Farrell, Diana, and Andrew Grant. 2005. "China's Looming Talent Shortage." *McKinsey Quarterly* 4:72.

Gallucci, Maria. 2016. "As US Oil Industry Jobs Decline, Undergraduates Are Ditching Petroleum Engineering Degrees." *Inside Business Times*, Mar. 18. http://www.ibtimes.com/us-oil-industry-jobs-decline-undergraduates-are-ditching-petroleum-engineering-2338662.

Hill, Daniel, and Stephen A. Holditch. 2013. "Guest Editorial: Déjà Vu All Over Again." *Journal of Petroleum Technology* 65 (6).

IEEE (Institute of Electrical and Electronics Engineers). 2011. "New IEEE-USA President Looks to Advance US Innovation and Entrepreneurship." *IEEE Antennas and Propagation Magazine* 53 (1): 196.

Isaacson, Walter. 2011. *Steve Jobs*. New York: Simon & Schuster.

Kemp, John. 2015. "Mass Layoffs Complicate Oil Industry's Long-Term Plans: Kemp." *Reuters*, Feb. 16. http://www.reuters.com/article/oil-employment-graduates-kemp-idUSL5N0VN2ZN20150213.

Lassetter, Susan 2014. "Mississippi State to Offer State's Only Petroleum Engineering Degree." Mississippi State University College of Engineering, Nov. 20. Accessed June 20, 2017. http://www.bagley.msstate.edu/news/mississippi-state-to-offer-states-only-petroleum-engineering-degree/.

Lowell, B. Lindsay, and Hal Salzman. 2007. "Into the Eye of the Storm: Assessing the Evidence on Science and Engineering Education, Quality, and Workforce Demand." Paper presented at the annual meeting of the Association for Public Policy Analysis and Management, Washington, D.C., Nov. 8–10.

Lynn, L., and H. Salzman. 2010. "The Globalization of Technology Development: Implications for U.S. Skills Policy." In *Transforming the U.S. Workforce Development System: Lessons from Research and Practice*, edited by David Finegold, Mary Gatta, Harold Salzman, and Susan J. Shurman, 57–86. New York: Labor and Employment Relations Association.

Marchand, Joseph. 2012. "Local Labor Market Impacts of Energy Boom-Bust in Western Canada." *Journal of Urban Economics* 71 (1): 165–74.

Margo, Robert A. 2000. *Wages and Labor Markets in the United States, 1820–1860*. Chicago: University of Chicago Press.

Meiksins, P., and C. Smith. 1996. *Engineering Labor: Technical Workers in Comparative Perspective*. London: Verso.

National Association of Colleges and Employers. 2009. *Salary Survey*, vol. 48, iss. 2. Bethlehem, PA: NACE.

———. 2010. *Salary Survey*, vol. 49, iss. 3. Bethlehem, PA: NACE.

———. 2014. *Salary Survey*. Bethlehem, PA.

National Association of Manufacturers. 2005. *The Looming Workforce Crisis: Preparing American Workers for 21st Century Competition*. Washington, D.C.: National Association of Manufacturers.

National Research Council. 2007. *Rising above the Gathering Storm: Energizing and Employing America for a Brighter Economic Future*. Washington, D.C.: National Academies Press.

———. 2010. *Rising above the Gathering Storm, Revisited: Rapidly Approaching Category 5*. Washington, D.C.: National Academies Press.

———. 2012. *Assuring DoD a Strong Science, Technology, Engineering, and Mathematics (STEM) Workforce*. Washington, D.C.: National Academies Press.

President's Council on Jobs and Competitiveness. 2011. "President's Council on Jobs and Competitiveness Announces Industry Leaders' Commitment to Double Engineering Internships in 2012." Press release, Aug. 31. https://obamawhitehouse.archives.gov/the-press-office/2011/08/31/president-s-council-jobs-and-competitiveness-announces-industry-leaders.

Princeton Review. 2013. "Career: Petroleum Engineer." Accessed Feb, 2013. http://www.princetonreview.com/careers.aspx?cid=109.

Rubin, Mark. 2010. "Workforce Dilemma Requires Action." *American Oil & Gas Reporter*, Mar. 20. http://www.aogr.com/index.php/magazine/editors_choice /march_2010_ed.

Ruggles, Steven, J. Trent Alexander, Katie Genadek, Ronald Goeken, Matthew B. Schroeder, and Matthew Sobek. 2017. Integrated Public Use Microdata Series: Version 7.0 [Machine-readable database]. Minneapolis: University of Minnesota.

Salzman, Hal. 2013. "What Shortages? The Real Evidence about the STEM Workforce." *Issues in Science and Technology* (National Academy of Science policy magazine) 29 (4). http://issues.org/29-4/what-shortages-the-real-evidence-about -the-stem-workforce/.

Salzman, Hal, and Lindsay Lowell. 2008. "Making the Grade." *Nature* 453:28–30.

Salzman, Hal, and Leonard Lynn. 2010. "Engineering and Engineering Skills: What's Really Needed for Global Competitiveness." Paper presented at the annual meeting of the Association for Public Policy Analysis and Management, Boston, MA, Nov. 4.

Saunders, Barbara. 2011. "The Great Crew Change: 'Wolf Cries' or Reality?" *Rigzone*, Aug. 12. Accessed Aug. 12, 2011. http://www.rigzone.com/news/article .asp?a_id=110134.

Shauk, Zain. 2015. "Is There Money to Be Made in Oil? New Grads Don't Think So." Accessed Jan. 28, 2015. http://www.bloomberg.com/news/articles/2015-01-29 /graduates-question-six-figure-oil-jobs-as-layoffs-swell.

Society of Petroleum Engineers (SPE). 2015. "Fewer Petroleum Engineering Graduates Finding Jobs in Industry." Accessed June 20, 2017. http://www.spe.org/news /article/fewer-petroleum-engineering-graduates-finding-jobs-in-industry.

Teitelbaum, Michael S. 2014. *Falling Behind? Boom, Bust, and the Global Race for Scientific Talent*. Princeton, NJ: Princeton University Press.

Weaver, Andrew. 2017. "Chemical and Petroleum Engineering: The Numbers." Mississippi State University College of Engineering. Accessed June 20, 2017. http:// www.che.msstate.edu/chemical-and-petroleum-engineering-the-numbers/.

Yoder, Brian L. 2013. "Engineers by the Numbers." American Society for Engineering Education. https://www.asee.org/papers-and-publications/publications /college-profiles/15EngineeringbytheNumbersPart1.pdf.

Bridge to Permanent Immigration or Temporary Labor?
The H-1B Visa Program Is a Source of Both

Ron Hira

9.1 Introduction

Many members of the popular press, pundits, business and university leaders, and policymakers conflate and often confuse guest worker visas, such as the H-1B, with permanent immigration.[1] Carly Fiorina, an advisor to John McCain's presidential campaign in 2008 and former chief executive officer (CEO) of Hewlett-Packard, responded to a question about H-1Bs during the campaign this way, "It is in our economic interest to have really smart people wanting to come here. And so what's wrong with the H-1B visa system today, among other things, is that we curtail that program so tightly that the limits that Congress allows for H-1B visa entrance are usually filled within one week. So we have to find a more practical system for allowing smart, hardworking people to come into this country and it should be our goal to get them to stay here forever" (Bomey 2008). Reading the quote, one might expect that expanding the H-1B program is the critical change to immigration policy that is needed in order to keep skilled workers here permanently. While permanent residence allows foreign nationals to live and work in the United States permanently, guest worker visas like an H-1B allow them to live and work in the United States only temporarily (not "forever") and under circumstances that restrict their ability to stay in the coun-

Ron Hira is associate professor of political science at Howard University.

For acknowledgments, sources of research support, and disclosure of the author's material financial relationships, if any, please see http://www.nber.org/chapters/c12691.ack.

1. Some justify expansion of the H-1B program on the grounds that immigrants found new companies in the United States (Friedman 2009; *Washington Post* 2008). However, by regulations H-1Bs are not allowed to found a company.

try if they do not keep their position. These circumstances sometimes put guest workers in a precarious position that invites their exploitation, creates insecurity for them, and undermines the integrity of the labor market. These consequences are caused by the design of immigration policies—a combination of loopholes and the fact that employers, rather than workers, control the work permit.

While some H-1B visa holders gain permanent residence, many employers use the H-1B program solely for employing temporary immigrants, and their share of the H-1B visa numbers is large and increasing. This chapter shows that the H-1B guest worker program has bifurcated, with some employers using the H-1B visa program as a bridge to permanent immigration while *most* top users of the H-1B visa programs sponsor very few of their workers for permanent residence. Firms that use the program principally to offshore work to lower-cost countries use H-1Bs as temporary labor. They pay lower wages, have flatter wage distributions, source a much higher share of their H-1Bs from India, and have a higher proportion of their H-1Bs with no more than a bachelor's degree compared to firms that sponsor H-1Bs for permanent residency. There are differences in H-1B use in sponsoring permanent residence even within different divisions of the same company. Given the relatively low wages that can be paid to H-1B visa holders, firms have increasingly used the program for temporary labor mobility to transfer work overseas and to take advantage of lower-cost guest worker labor rather than attracting the "best and brightest" for permanent immigration. High-skilled immigration policy discussions should recognize these empirical realities.

9.1.1   Permanent Residence versus Guest Worker Status

The distinction between a permanent residence visa, commonly called a green card, and guest worker status is substantial and has important economic and policy implications, particularly for the high-skilled labor market (and especially in the information technology and engineering labor markets). Permanent residents enjoy similar employment rights as American citizens. They are eligible to apply for nearly all the same jobs as citizens, and they can stay in the United States even if they are out of the labor market. On the other hand, H-1B visas are work permits held by a specific employer for up to six years. The employer holds the work permit so it can revoke the visa at any time by terminating the worker, which means that the worker must leave the country immediately.[2] The H-1B workers can switch employers only if they can find another employer willing to sponsor them for an H-1B. In contrast to the employment rights of citizens and permanent residents, H-1B rules place most of the power in the hands of the employer

---

2. Generally, workers who are laid off try to switch status to a nonwork temporary visa, such as a tourist visa, while they search for work.

and create opportunities for leverage that allows some employers to exploit guest workers for whom they obtained an H-1B. Some, such as former Secretary of Labor Ray Marshall, have described this employment relationship as indentured (Marshall 2009, 37).

This type of exploitation has been widely reported in the press. A 2009 *Businessweek* cover story profiling the exploitation of H-1B workers was called, "America's High-Tech Sweatshops" (Hamm and Herbst 2009). Also in 2009, the Louisiana Federation of Teachers filed a complaint on behalf of teachers brought in from the Philippines, who were being held in "virtual servitude." Their employer intimidated them, charged exorbitant and unnecessary fees, and forced them to live in roach-infested, run-down apartments leased by the employer (Toppo and Fernandez 2009). This type of exploitation is not new. Back in 1993, CBS's *60 Minutes* television show aired a story on H-1B computer programmers who were contracted out to Hewlett-Packard for $10 per hour, nowhere near what the company would have to pay permanent residents.[3]

Current U.S. immigration policy favors family-based immigration, which accounts for about 65 percent of the approximately one million new permanent immigrants admitted annually. Many skilled immigrants come through family-based immigration, but H-1B visas serve as important sources of skilled permanent immigration. A majority of permanent, employment-based immigrants were originally H-1Bs. The visas are "dual intent," meaning that while visa holders are here temporarily on nonimmigrant work permits, their status does not preclude them from staying permanently if their employer chooses to apply for an employment-based permanent immigration visa. Employment-based immigration accounts for approximately 15 percent of permanent immigration, and some researchers estimate that 62 percent of employment-based permanent immigrants began as H-1B temporary workers (Jasso et al. 2010). To say that the H-1B accounts for a majority of employment-based permanent immigration is not, however, the same as saying that most H-1Bs become permanent residents. Many H-1B workers are never sponsored for permanent residence. The H-1B workers cannot sponsor themselves for permanent immigration. Only employers have that authority and exercise it at their discretion. For those guest workers who want to stay permanently, it puts additional power in the hands of their employers, power that employers have lobbied to maintain. During the 2007 debate over comprehensive immigration reform, businesses fought against an allocation of self-sponsored, high-skill immigrant visas based on a merit point system, arguing that they, as employers know best what kind of workers are needed as permanent residents in the United States (Hennessy-Fiske and Puzzanghera 2007).

3. CBS television broadcast, Oct. 3, 1993.

9.1.2   H-1B Visas: Preimmigration versus Temporary Worker

The H-1B is a nonimmigrant visa under the Immigration and Nationality Act (INA), section 101(a)(15)(H). It allows employers within the United States to temporarily employ foreign workers in *specialty occupations*. The regulations define a "specialty occupation" as requiring theoretical and practical application of a body of highly specialized knowledge in a field of human endeavor including, but not limited to, architecture, engineering, mathematics, physical sciences, social sciences, biotechnology, medicine and health, education, law, accounting, business specialties, theology, and the arts, and requiring, with the exception of fashion models, the attainment of a bachelor's degree or its equivalent as a minimum. Likewise, the foreign worker must possess at least a bachelor's degree or its equivalent and state licensure if required to practice in that field. The H-1B work authorization is strictly limited to employment by the sponsoring employer. The duration of the visa is three years, extendable to a maximum of six. However, this duration can be extended indefinitely beyond the six years, in one-year increments, if the employer is sponsoring the H-1B worker for permanent residence.

9.1.3   The Data

The H-1B is a large guest worker program, admitting 124,326 *new* foreign workers in fiscal year 2014 alone (U.S. Department of Homeland Security 2015). While no one knows the exact number of H-1B holders in the United States at any one time, because the government does not track those numbers, estimates are in the range of 650,000. For the H-1B data, I use the I-129 petitions approved by USCIS, which I received via a Freedom of Information Act (FOIA) request in 2013. The data set covers all approved petitions for fiscal years 2010–2012. Each petition is for an individual worker and includes the name of the employer as well as specific characteristics of the worker such as wages, highest education level attained, and worker's country of origin. The data set was cleaned to correct for firm misspellings and to consolidate firm subsidiaries.

A nonimmigrant visa can be an important first step toward permanent residence for many skilled foreign workers, but most never become permanent residents. Even before the emergence of the offshoring of high-skill jobs, many H-1Bs were never converted to permanent residence by employers. Lowell (2000) estimated that at its peak 47 percent of H-1Bs became permanent residents. To analyze this process more closely I estimate permanent sponsorship rates by employer for the top twenty H-1B firms. I do this by using the Program Electronic Review Management (PERM) database, which is kept by the U.S. Department of Labor's Office of Foreign Labor Certification.[4]

4. The data can be found from the U.S. Department of Labor here: https://www.foreignlabor cert.doleta.gov/performancedata.cfm#dis.

Employment-based immigration is a four-step process. The first step, sometimes called pre-PERM, is for the employer to complete active recruitment of U.S. workers by advertising in newspapers and collecting applications.[5] Once the recruitment takes place, and presumably the employer has not found a qualified American applicant, the employer files an "Application for Permanent Employment Certification" (ETA Form 9089) with the U.S. Department of Labor. The data for each of these cases are entered into the PERM database. I have combined the FY2010, 2011, and 2012 data sets. According to the PERM database (the U.S. Department of Labor's Permanent Labor Certification Program Database), H-1Bs accounted for 77 percent of the permanent residence applications, or 142,695 of the 184,682, in the three-year period FY2010–12. So, it is clear that a large share of the PERM applications is for workers in an H-1B visa status.

9.1.4   Different H-1B Uses: Preimmigration versus Way Station

As mentioned earlier, different employers use the H-1B program either as a bridge to permanent immigration or as a temporary labor mobility program. Even within different divisions of the same company, employers will use its guest worker visas differently—some divisions use it for a conversion to permanent residence while other divisions use it purely for temporary labor mobility. An exemplary case of this divergence is Silicon Valley–based software giant Oracle Corporation. When asked whether Oracle uses the H-1B program as a bridge to immigration, Robert Hoffman, then lobbyist and vice president for government affairs at Oracle, stated, "More than 90 percent of Oracle's visa workers are trying to stay in the United States and are on the path to permanent residency" (McGee 2007). At nearly the same time as Hoffman's statement, Shahab Alam, an executive of I-Flex (now known as Oracle Financial Solutions), a subsidiary of Oracle, described its use of the H-1B visas as unrelated to permanent residency (NPR Marketplace Radio 2007): "Most of the people coming through us [on H-1B] have no intention of settling in the United States. These are folks who are coming here to do a job, have fun while they can in the United States, and then use this experience in different parts of the world."[6]

The government does not directly measure the conversion from temporary to permanent resident, but we can use available data to estimate

5. A number of serious weaknesses in this process have been identified, where firms go through the motions of recruitment with the goal of excluding qualified American workers from being hired. This process was described in a video made by the immigration law firm, Cohen & Grigsby, in a marketing seminar. The video became viral in 2007 and excerpts can be seen here: http://www.youtube.com/watch?v=TCbFEgFajGU. American worker groups like the Programmers Guild have complained repeatedly about what they describe as "fake PERM ads," where these ads are not bona fide job opportunities.

6. This contrast between Oracle and I-Flex is particularly interesting because, at the time, Robert Hoffman served as the chief spokesperson for Compete America, the primary business and educational coalition lobbying for H-1B increases. Given the significant use of H-1Bs by I-Flex, the only way Hoffman could be faithfully reporting Oracle's use was by excluding I-Flex's numbers in his calculations. In fact, in FY2007, when both of these interviews took

it. To examine this "bridge to immigration" I introduce a measure I call *immigration yield*, which is the ratio of *PERM applications filed for H-1B workers* to *initial H-1B petitions received* by a specific employer. As noted above, detailed PERM applications are available from the U.S. Department of Labor's Office of Foreign Labor Certification. Beginning in 2007, the PERM data included the current visa status (H-1B, L-1, O-1, E-3, etc.) for each employee, so one can calculate the yield for each visa type. Ideally, we would be able to track each individual guest worker to identify whether they are sponsored for, and later granted, permanent residence, but names of workers are considered private and therefore not released in either the H-1B I-129 data nor in the PERM data. The data presented below should be viewed as *indicators* of the conversion rates for different employers rather than as literal rates. An employer could wait a number of years before beginning the PERM process for its guest workers, so that workers who are not being sponsored at one period could be sponsored at a future time. Another reason is that even after an employer initiates the process for converting a guest worker from an H-1B there is a lead time before the application appears in the PERM database. The lead times are due to regulatory requirements such as advertising the position in newspapers to search for American workers and for Department of Labor processing. To mitigate these effects I use a three-year period FY2010–12 instead of just a single year. Last, there are so-called priority workers of extraordinary ability or multinational executives or university professors, who are sponsored on EB-1 permanent visas. Those workers are not subject to the labor certification, so their employer can bypass the form that populates the PERM database. In FY2012, EB-1s accounted for 16,286 of the employment-based permanent residences granted, and the majority, 9,209, were for multinational executives on L-1A, a different guest worker visa. The small numbers of H-1Bs who are sponsored through an EB-1 are not likely to bias the conversion rates discussed below. Notwithstanding these limitations the data show clear and distinctive patterns of H-1B use by employers: some employers use it for purely temporary purposes while others use it as a bridge to permanent immigration.

9.2 H-1B Use by Offshoring Firms versus Product Firms

Table 9.1 shows the immigration yields for the top twenty H-1B employers for the three-year period FY2010–12. The top twenty H-1B employers account for a large share of the FY2010–12 visas issued. These visas are

place, I-Flex received 374 H-1Bs but applied for permanent residence for only sixteen of its H-1B workers, an immigration yield of only 4 percent. That is a far cry from the 90 percent Hoffman claimed. And in 2007, I-Flex received more than three times as many H-1Bs as its parent, Oracle, which received 113 worker visas used for intracompany transfers (U.S. Department of Homeland Security 2012).

Table 9.1   FY2010–12 top twenty H-1B employers: Immigration yield

H-1B rank	Firm	FY2010–12 new H-1Bs received	FY2010–12 PERM applications for H-1B workers	Immigration yield (%)	Significant offshoring
1	Cognizant	17,964	2,228	12	X
2	Tata Consultancy Services	9,083	—	0	X
3	Wipro	8,726	98	1	X
4	Infosys	6,550	129	2	X
5	Accenture	5,799	27	0	X
6	Microsoft	4,766	4,265	89	
7	IBM	3,770	462	12	X
8	Larsen & Toubro	3,286	50	2	X
9	HCL	3,074	276	9	X
10	Deloitte	2,850	591	21	X
11	Mahindra Satyam	2,535	41	2	X
12	Intel Corp	2,036	917	45	
13	Patni-iGate	1,960	186	9	X
14	Syntel	1,646	53	3	X
15	Google	1,477	705	48	
16	Amazon	1,378	614	45	
17	Qualcomm	1,265	1,247	99	
18	PricewaterhouseCoopers	1,059	392	37	X
19	Mphasis	993	106	11	X
20	Synechron	700	26	4	X
	Total	80,917	12,413	15	15 of 20

Sources: H-1B data from USCIS I-129 petitions; PERM data from U.S. Department of Labor, Foreign Labor Certification Data Center.

capped with an annual quota of 85,000. The top twenty H-1B employers received 80,917, or 32 percent, of the three years' worth of 255,000 allotted. These firms are a significant determinant of the impact of the H-1B program on the U.S. economy and labor market. The H-1B program is employer driven and employers have considerable discretion over their use of the program. The employer selects which foreign workers to hire as an H-1B and which ones to sponsor for permanent immigration. As a result, firm behavior is the significant driver of program impact. How do employers use the H-1B program in relation to permanent immigration? Which employers use the H-1B program as a bridge to permanent immigration versus those using it for temporary labor? Do H-1B worker characteristics vary across these employers?

What emerges from the analysis is employer use clustered around two business models. The first model is an *offshoring* business model, in which companies perform most of their work overseas in low-cost countries. *Off-shoring firms* in information technology (IT) include firms that outsource most of their work overseas such as Cognizant, Infosys, Wipro, and Tata Consultancy Services and major IT and consulting firms like Accenture, Deloitte, and IBM that have built up significant offshore outsourcing operations. These companies perform most of their work overseas in low-cost countries.[7] The second business model, in which firms do considerable work in the United States, are *product* firms that do not provide significant offshoring outsourcing, such as Intel, Qualcomm, and Microsoft.

In the rest of this chapter I contrast these two types of firms along several dimensions by giving statistics on them separately in a single table, with an "A" name for the offshoring firms and a "B" name for the product firms. Thus, tables 9.2A, 9.3A, 9.4A, and 9.5A all refer to the data from the same firms, those I characterize as offshoring firms, while tables 9.2B, 9.3B, 9.4B, and 9.5B all refer to data for product firms.

9.3 Offshoring Firms Have Low Immigration Yields While Product Firms Have High Yields

The H-1B visa rankings in table 9.2A show very low immigration yield for most of the major offshore outsourcing firms for FY2010–12, indicating that these firms have little interest in converting their H-1B employees to permanent residence. These firms are the largest users of the H-1B program, making up fifteen of the top twenty users. Those fifteen firms alone received nearly 70,000 visas, or 86 percent of the top twenty. The list here also mirrors the largest of the offshore outsourcing firms. The business model of these firms is to transfer labor overseas—not to hire in the United States permanently. In fact, many of these firms hire very few American citizens and, as their immigration yields show, sponsor few H-1Bs for permanent residence (Srivastava and Herbst 2010). Tata Consultancy Services, the largest India-based offshore outsourcing firm, did not file an application for a single permanent resident for any of its H-1Bs. The pure-play offshore outsourcing firms all have immigration yields at 12 percent or below. Cognizant, the top firm, which is headquartered in the United States, has the highest immigration yield in the group at 12 percent.[8] IBM is a hybrid firm, with business segments beyond offshoring that include product lines of semiconductors and packaged software. This analysis combines IBM with IBM India, a wholly owned subsidiary of U.S.-based IBM (IBM 2008, exhibit 21). IBM

---

7. See my policy brief for immigration for the Agenda for Shared Prosperity (Hira 2007), and for a more detailed treatment of the offshore outsourcing phenomenon, see my book, *Outsourcing America* (Hira and Hira 2008).

8. Even though Cognizant, a spin-off of Dun & Bradstreet, is based in the United States, its business model is the same as the India-based offshore outsourcing firms. Cognizant's CEO Lakshmi Narayanan served as the chairman of NASSCOM (the Indian industry association for offshore outsourcing) in 2007.

Table 9.2A   **Offshoring firms: Immigration yield**

H-1B rank	Firm	FY2010–12 new H-1Bs received	FY2010–12 PERM applications for H-1B workers	Immigration yield (%)	Pure-play offshoring firm
1	Cognizant	17,964	2,228	12	X
2	Tata Consultancy Services	9,083	—	0	X
3	Wipro	8,726	98	1	X
4	Infosys	6,550	129	2	X
5	Accenture	5,799	27	0	
7	IBM	3,770	462	12	
8	Larsen & Toubro	3,286	50	2	X
9	HCL	3,074	276	9	X
10	Deloitte	2,850	591	21	
11	Mahindra Satyam	2,535	41	2	X
13	Patni-iGate	1,960	186	9	X
14	Syntel	1,646	53	3	X
18	PricewaterhouseCoopers	1,059	392	37	
19	Mphasis	993	106	11	X
20	Synechron	700	26	4	X

Sources: H-1B data from USCIS I-129 petitions; PERM data from U.S. Department of Labor, Foreign Labor Certification Data Center.

India applied for zero PERMs. Its operations are similar to the other pure-play offshore outsourcing firms, and IBM identifies Wipro and Satyam as IBM India's competitors in its annual report (IBM 2008). Two outliers in this list are Deloitte and PricewaterhouseCoopers, with immigration yields of 21 percent and 37 percent, respectively. Both firms are part of the "Big Four" in the accounting industry. They use the H-1B program for multiple purposes that vary across the different business lines. Both have core tax and audit business lines, which use the H-1B program mostly as a bridge to permanent residence, while their consulting arms, which compete directly with Accenture and IBM in the offshore outsourcing sector, use the H-1B for temporary mobility. It is clear that the offshoring firms have little or no interest in sponsoring their H-1B workers for permanent residence, and some have been quite clear about it publicly.

Most of the firms in table 9.2A are members of NASSCOM, India's offshore outsourcing trade association. While he served as president of NASSCOM, Som Mittal, a former executive of Hewlett-Packard India, described why the H-1B program is so important to his member firms, "We need for people to travel back and forth between the United States and India to consult on and complete projects" (Herbst 2009). NASSCOM and the Indian government view the H-1B visa as trade, rather than immigration, policy issues. They believe that their primary comparative advantage is low-cost, high-skilled workers, and that H-1B regulations, such as wage floors and quotas, are nontariff barriers to trade.

9.3.1   Why Not Hire American Workers?

How do low immigration yields for offshore outsourcing firms fit into their human resource practices and what does it tell us about their business models? Offshore outsourcing firms have demonstrated little interest in hiring American workers. The business model is reselling labor, and the H-1B workers can be paid less than an American worker. For example, even though Tata Consultancy had more than 10,843 workers in the United States in 2007, only 739 (9 percent) were Americans. Why are these firms not interested in hiring American workers?

Offshore outsourcing firms rely on the H-1B programs for three principal reasons.

First, it facilitates their knowledge-transfer operations, where they rotate in foreign workers to learn U.S. workers' jobs.

Second, the H-1B program provides them an inexpensive, on-site presence that enables them to coordinate offshore functions. Many functions that are done remotely still require a significant amount of physical presence at the customer site. For example, according to its financial reporting, Infosys's on-site workers, almost all of whom are foreign guest workers, directly accounts for slightly less than half of its overall revenue (Infosys 2009, slide 12). And according to a Tata Consultancy Services executive, H-1B workers are less expensive than comparable American workers. Then vice president Phiroz Vandrevala described, in an interview with an India-based business magazine, how his company derives competitive advantages by paying its visa holders below-market wages: "Our wage per employee is 20–25 percent lesser than U.S. wage for a similar employee," Vandrevala said. "Typically, for a TCS employee with five years' experience, the annual cost to the company is $60,000–70,000, while a local American employee might cost $80,000–100,000. This (labour arbitrage) is a fact of doing work onsite. It's a fact that Indian IT companies have an advantage here and there's nothing wrong in that. . . . The issue is that of getting workers in the U.S. on wages far lower than local wage rate" (Singh 2003). Neeraj Gupta and Brian Keane, veterans of the IT services industry say that the H-1B program allows IT services firms to save 20 percent to 25 percent by hiring an H-1B worker over hiring an American one.[9]

Third, the H-1B program allows the U.S. operations to serve as a training ground for foreign workers who then rotate back to their home country to do the work more effectively than they could have without such training in the United States. A *Businessweek* story quoted an executive from Wipro

---

9. Author conversations with Brian Keane and Neeraj Gupta on March 14, 2013.

describing the company's use of the H-1B program: "Wipro has more than 4,000 employees in the United States, and roughly 2,500 are on H-1B visas. About 1,000 new temporary workers come to the country each year, while 1,000 rotate back to India, with improved skills to serve clients" (Elstrom 2007). Some firms use the H-1B visas for *knowledge transfer*, where an incumbent worker transfers his detailed knowledge about his jobs to the trainee, with the explicit purpose of laying off their higher-cost American workers. Firms sometimes do the replacement through contractors.

An example of this behavior in 2003 gained congressional attention and was the centerpiece of a number of congressional hearings. In Lake Mary, Florida, Siemens used Tata Consultancy Services to replace its American workers with guest workers earning one-third of the wages. In an award-winning series, business reporter Lee Howard of *The Day* newspaper documented how Pfizer was forcing its U.S. workers to train foreign replacements from offshore outsourcers Infosys and Satyam (Howard 2008). In another example, the television ratings firm Nielsen forced its American workers to train foreign replacements working for Tata Consultancy Services. This took place while Nielsen received tax incentives from local government to create jobs (Kruse and Blackwell 2008). And in 2009, workers at Wachovia, which was being bailed out by the government through the Troubled Asset Relief Program (TARP), claimed they were training their foreign replacements on H-1B visas (Bradley 2009).

Southern California Edison's replacement of 500 of its American IT workers with H-1B workers employed at Tata and Infosys was profiled during a U.S. Senate hearing (Grassley 2015). Southern California Edison told its American IT workers that it was replacing them because the H-1B workers are cheaper. The wage differentials are stark, with the American IT workers earning $110,000 while the H-1B workers replacing them are earning $65,000 to $70,000 (Hira 2015).

The H-1B visas are vital to the scalability of the offshoring business model, so some firms are "banking" visas, that is, keeping excess H-1B workers in their home countries and sending them to the United States only as the need arises. The firms measure their slack H-1B visas in terms of utilization rates; that is, the percent of their H-1Bs are actively in the United States. During an earnings call with Wall Street research analysts covering the firm, Infosys's chief operating officer Kris Gopalkrishnan responded to questions about whether it has an adequate number of workers with visas by saying, "It is 37% of the total visas available right now with Infosys [that are] being used. That means we have remaining 63% of the people having visas available to put on projects. So it gives us a better utilization rate or—so it gives us the flexibility. We typically get worried when it reaches 50%–55% because that means that we may not be able to find the right people with the visas two [*sic*] deploy on the project, so 37% is a comfortable number" (Infosys 2005).

These guest worker visas are so integral to the offshore outsourcing firms

Table 9.2B   **Product firms: Immigration yield**

H-1B rank	Firm	FY2010–12 new H-1Bs received	FY2010–12 PERM applications for H-1B workers	Immigration yield (%)
6	Microsoft	4,766	4,265	89
12	Intel Corp.	2,036	917	45
15	Google	1,477	705	48
16	Amazon	1,378	614	45
17	Qualcomm	1,265	1,247	99

Sources: H-1B data from USCIS I-129 petitions; PERM data from U.S. Department of Labor, Foreign Labor Certification Data Center.

that then-Indian Commerce Minister Kamal Nath called the H-1B the "outsourcing visa" in an interview with the *New York Times* while arguing for an increase in the H-1B cap (Lohr 2007). In responding to the competitive threat from offshore outsourcing firms like Infosys, many multinational corporations, which until recently have had traditional business models, have moved very aggressively to adopt their own offshore outsourcing business model. The primary business model of these firms is not offshore outsourcing, but they have built up significant offshore outsourcing operations. Some of these firms, such as Hewlett-Packard (HP), have done this through acquisitions (HP acquired EDS and MPhasis), or through subsidiaries, while others have simply transferred work to new employees in low cost countries.

Accenture and IBM provide interesting cases. Accenture has built up its workforce in low-cost countries very quickly. According to its CEO, as of August 2007, Accenture had more employees in India than any other country, including the United States (Chatterjee 2007). Similarly, IBM has increased its workforce in India very dramatically. From a mere 6,000 workers in India in 2003, its headcount rose to 74,000 by 2007, and is projected to have reached 100,000 by 2010 (D'Souza 2008; McDougall 2006). Given the continuing downsizing of its U.S. workforce, reduced to 115,000 in 2009, India likely became its largest workforce in 2012 (Lohr 2009).

Table 9.2B shows that product firms, which are not in the business of offshore outsourcing, are clustered into two groups with respect to their immigration yields. First, there are firms like Microsoft or Qualcomm that are heavy users of the H-1B and try to convert a large share of them to permanent residence. Then there is a group, Google, Intel, and Amazon where employers are converting about one-half of their H-1Bs to permanent residence.

9.4   Other Differences in H-1B Use between Offshoring and Product Firms

Tables 9.3A and 9.3B show substantial differences in the wage distributions of offshoring firms versus product firms use of the H-1B program.

Table 9.3A   **Offshoring firms: Wage distribution**

H-1B rank	Firm	FY2010–12 new H-1Bs received	5th percentile ($)	10th percentile ($)	25th percentile ($)	Median ($)	75th percentile ($)	90th percentile ($)	95th percentile ($)
1	Cognizant	17,964	51,000	53,100	57,100	61,197	70,500	77,000	85,483
2	Tata Consultancy Services	9,083	58,000	61,200	61,800	64,900	65,700	66,480	66,900
3	Wipro	8,726	60,000	60,000	60,000	64,854	71,406	78,136	85,946
4	Infosys	6,550	60,000	60,000	60,000	60,000	60,000	71,822	78,811
5	Accenture	5,799	48,600	52,900	58,500	64,700	70,100	81,300	88,900
7	IBM	3,770	58,200	60,000	64,200	70,500	80,205	100,000	115,000
8	Larsen & Toubro	3,286	44,700	46,860	51,460	56,226	60,268	65,273	69,868
9	HCL	3,074	51,854	55,643	60,000	61,000	68,100	76,870	84,083
10	Deloitte	2,850	50,000	54,960	61,526	68,500	80,000	105,000	130,000
11	Mahindra Satyam	2,535	60,000	60,000	60,000	62,400	68,109	75,629	79,102
13	Patni-iGate	1,960	46,790	48,800	55,600	62,900	70,100	76,400	79,525
14	Syntel	1,646	54,000	54,000	54,000	54,000	62,000	70,000	75,347
18	PricewaterhouseCoopers	1,059	50,000	51,500	55,000	60,000	75,000	100,000	120,000
19	Mphasis	993	60,000	60,000	60,000	62,130	67,870	76,353	80,475
20	Synechron	700	61,400	62,000	65,200	68,500	72,000	76,720	81,634

Source: USCIS I-129 petitions.

Table 9.3B   Product firms: Wage distribution

H-1B rank	Firm	FY2010–12 new H-1Bs received	5th percentile ($)	10th percentile ($)	25th percentile ($)	Median ($)	75th percentile ($)	90th percentile ($)	95th percentile ($)
6	Microsoft	4,766	78,000	80,000	81,000	95,000	104,000	120,000	130,661
12	Intel Corp.	2,036	72,400	73,800	77,392	84,976	100,000	105,400	113,100
15	Google	1,477	81,800	88,000	100,000	110,000	127,000	135,000	150,000
16	Amazon	1,378	80,000	87,000	90,000	95,000	100,000	115,000	120,000
17	Qualcomm	1,265	77,151	80,018	82,493	85,010	102,856	115,003	125,008

Source: USCIS I-129 petitions.

The firms with significant offshoring in table 9.3A all have lower absolute levels of wages, with medians ranging from $54,000 to $70,500, compared to the product firms' median wages in table 9.3B that range from $85,000 to $110,000.

Some pure-play offshore outsourcing firms have very flat wage distributions. For example, Infosys's wage at the 75th percentile is $60,000 and is exactly the same as its wage at the 5th percentile. That means almost 5,000 of Infosys's H-1B workers are paid exactly $60,000. This is likely due to the fact that H-1B regulations are more stringent for heavy users of H-1B firms (so-called H-1B dependent) that pay workers less than $60,000. The H-1B-dependent firms must perform active recruitment and adhere to non-displacement requirements unless they pay H-1B workers at least $60,000. Infosys is able to achieve regulatory relief by paying at least $60,000, but it has little incentive to pay more than that.

Tables 9.4A and 9.4B show a striking difference in the source countries of H-1Bs for the offshoring versus product firms. With the exception of PricewaterhouseCoopers and Deloitte, all of the offshoring firms have more than 90 percent of their H-1Bs come from India. For some, like U.S.-based firms Cognizant and Syntel, virtually all of their H-1Bs are from India. For the product firms, India is still the top source country for H-1Bs, but it isn't nearly as dominant. In the case of Google, China is the top source country for its H-1Bs. The product firm reliance on H-1B workers from India means that many of their workers will be waiting in long lines for permanent residency. Unlike the H-1B program, which has no per-country limits, there are

Table 9.4A   Offshoring firms: Top source country for H-1B workers

H-1B rank	Firm	FY2010–12 new H-1Bs received	Top source country	Number	Top source country share of total (%)
1	Cognizant	17,964	India	17,898	100
2	Tata Consultancy Services	9,083	India	9,057	100
3	Wipro	8,726	India	8,687	100
4	Infosys	6,550	India	6,341	97
5	Accenture	5,799	India	5,503	95
7	IBM	3,770	India	3,420	91
8	Larsen & Toubro	3,286	India	3,275	100
9	HCL	3,074	India	3,048	99
10	Deloitte	2,850	India	1,981	70
11	Mahindra Satyam	2,535	India	2,524	100
13	Patni-iGate	1,960	India	1,943	99
14	Syntel	1,646	India	1,642	100
18	PricewaterhouseCoopers	1,059	India	318	30
19	Mphasis	993	India	989	100
20	Synechron	700	India	692	99

Source: USCIS I-129 petitions.

Table 9.4B   Product firms: Top source country for H-1B workers

H-1B rank	Firm	FY2010–12 new H-1Bs received	Top source country	Number	Top source country share of total (%)
6	Microsoft	4,766	India	1,382	29
12	Intel Corp.	2,036	India	1,354	67
15	Google	1,477	China, People's Republic of	321	22
16	Amazon	1,378	India	644	47
17	Qualcomm	1,265	India	726	57

Source: USCIS I-129 petitions.

country-specific limits within the legal permanent resident quotas. In the case of India, as of February 2015 the backlog times range from ten years for advanced-degree holders and twelve years for those with no more than a bachelor's degree (U.S. State Department 2015, employment-based table).

Tables 9.5A and 9.5B show the H-1B beneficiary's highest level of education, bachelor's degree, master's degree, or doctorate. For the offshoring firms, bachelor's degree is the largest share of its H-1B workforce. For Infosys it accounts for 86 percent. Further, these firms hire virtually no doctorate holders, with IBM being the sole exception. As explained earlier, IBM is a hybrid company with business lines in offshoring as well as products like semiconductors and software. For the product firms, more than half of the H-1B workers for Amazon and Microsoft hold no more than a bachelor's degree. Intel, Google, and Qualcomm all hire some doctorate holders, with nearly one-third of Intel's H-1B workers holding a doctorate. The relatively low level of educational attainment is particularly surprising since much of the public discussion over H-1Bs presents them as recent advanced-degree graduates of U.S. universities. The educational bar for American workers and students to fill these positions is much lower than is widely believed.

9.5   Conclusion

To better understand the impacts of the H-1B program on the U.S. economy and labor market as well as for immigration policy, analysts need to examine the very different ways in which firms use the program. This chapter identified those firms that use the H-1B program as a bridge to permanent immigration versus those that are using it for temporary labor mobility. Among the top twenty H-1B employers, offshoring firms sponsor few, if any, of its H-1B workers for permanent residency while product firms tend to sponsor at higher rates. Further, among the top twenty H-1B employers, offshoring firms tend to pay lower wages, have a flatter wage distribution, and hire H-1B workers with lower levels of educational attainment. And offshoring firms rely on H-1B workers from India at the virtual exclusion

Table 9.5A   **Offshoring firms: Highest level of education for H-1B workers**

H-1B rank	Firm	FY2010–12 new H-1Bs received	Highest level of ed. is BS	BS share of total (%)	Highest level of ed. is MS	MS share of total (%)	Highest level of ed. is PhD	PhD share of total (%)
1	Cognizant	17,964	14,467	81	3,486	19	2	0
2	Tata Consultancy Services	9,083	7,053	78	2,023	22	1	0
3	Wipro	8,726	5,510	63	3,057	35	5	0
4	Infosys	6,550	5,613	86	905	14	4	0
5	Accenture	5,799	4,221	73	1,565	27	7	0
7	IBM	3,770	2,253	60	1,273	34	226	6
8	Larsen & Toubro	3,286	2,756	84	530	16		0
9	HCL	3,074	1,759	57	1,289	42	3	0
10	Deloitte	2,850	1,964	69	866	30	6	0
11	Mahindra Satyam	2,535	1,703	67	816	32	2	0
13	Patni-iGate	1,960	1,487	76	465	24	3	0
14	Syntel	1,646	1,172	71	472	29		0
18	PricewaterhouseCoopers	1,059	668	63	374	35	6	1
19	Mphasis	993	642	65	340	34		0
20	Synechron	700	419	60	277	40		0

Source: USCIS I-129 petitions.

Table 9.5B   Product firms: Highest level of education for H-1B workers

H-1B rank	Firm	FY2010–12 new H-1Bs received	Highest level of ed. is BS	BS share of total (%)	Highest level of ed. is MS	MS share of total (%)	Highest level of ed. is PhD	PhD share of total (%)
6	Microsoft	4,766	2,966	62	1,564	33	213	4
12	Intel Corp.	2,036	156	8	1,231	60	640	31
15	Google	1,477	650	44	668	45	148	10
16	Amazon	1,378	730	53	584	42	60	4
17	Qualcomm	1,265	444	35	711	56	109	9

Source: USCIS I-129 petitions.

of workers from any other country. Further analysis of the H-1B data at the firm and industry level, using I-129 microdata, can shed light about program impacts and provide policymakers with a better understanding about how to craft policy changes.

By design, current high-skill immigration policies in the United States place enormous power in the hands of employers. Employers hold the H-1B visa for workers, and employers have complete discretion whether and when to apply for permanent residence for those workers. There are very long backlogs for employment-based immigration for workers from particular countries, such as India. Once an employer applies for permanent residence for the worker, that worker cannot change jobs within the company, even to take a promotion, without hurting his chances for a green card (Ferriss 2006). If a worker who is being sponsored for a green card decides to change jobs, he would have to go to the back of the green card queue. This means that H-1B workers being sponsored for green cards are essentially tethered to their specific employer for very long periods of time. This reduces the worker's bargaining power and it also negatively affects technological innovation by restricting the movement of workers between employers.

The very large numbers of H-1B workers, coupled with the smaller allotment of employment-based immigration visas, often put guest workers who want to become permanent residents in a state of indentured limbo. The public policy discussion about high-skilled immigration has largely ignored the differences between guest worker visas, like the H-1B, and permanent residence. New policy designs should take into account these differences as well as how the two programs are connected to one another.

References

Bomey, Nathan. 2008. "H-P Ex-CEO Carly Fiorina: Cut Taxes, Retrain Workers." *Michigan Business Review*, May 29. Accessed May 29, 2008. http://www.mlive .com/rebrandingmichigan/index.ssf/2008/05/hp_exceo_carly_fiorina_cut_tax .html.

Bradley, Jim. 2009. "Foreign Workers Could Be Replacing Charlotte Bank Employees." *WSOC TV.com*, Mar. 31. Accessed on Dec. 13, 2009. http://www.wsoctv.com /news/19047187/detail.html.

Chatterjee, Sumeet. 2007. "Accenture to Raise India Staff to 35,000 by August." *Reuters*, Jan. 29.

D'Souza, Savio. 2008. "IBM Says Double-Digit Sales Growth in India to Stay." *Reuters*, Nov. 26.

Elstrom, Peter. 2007. "Work Visas May Work against the U.S." *Businessweek*, Feb. 8.

Ferriss, Susan. 2006. "Hired Hands: Wait for Green Card Tries Visa Holders." *Sacramento Bee*, Aug. 3.

Friedman, Thomas. 2009. "Open Door Bailout." *New York Times*, Feb. 10.

Grassley, Charles. 2015. Prepared Statement by Senator Chuck Grassley of Iowa,

Chairman, Senate Judiciary Committee, at a hearing titled "Immigration Reforms Needed to Protect Skilled American Workers." Mar. 17. Accessed Apr. 8, 2015. http://www.judiciary.senate.gov/imo/media/doc/03-17-15%20Grassley%20State ment1.pdf.

Hamm, Steve, and Moira Herbst. 2009. "America's High-Tech Sweatshops." *Businessweek*, Oct. 1.

Hennessy-Fiske, Molly, and Jim Puzzanghera. 2007. "Immigration Plan Doesn't Add Up, Critics Say: Businesses Fault the Senate Bill's Point System, Saying It Can't Keep Pace with the Changing Economy." *LA Times*, May 24. Accessed Dec. 13, 2009. http://articles.latimes.com/2007/may/24/nation/na-points24.

Herbst, Moira. 2009. "The H-1B Lull Is Only Temporary." *Businessweek*, Nov. 2. Accessed Nov. 2, 2009. https://www.bloomberg.com/news/articles/2009-11-02 /the-h-1b-visa-lull-is-only-temporary.

Hira, Ron. 2007. "Outsourcing America's Technology and Knowledge Jobs: High-Skill Guest Worker Visas are Currently Hurting Rather than Helping Keep Jobs at Home." In *EPI Agenda for Shared Prosperity Briefing Paper*. Washington, D.C.: Economic Policy Institute.

———. 2015. Testimony Given in a Hearing before the Judiciary Committee U.S. Senate on "Immigration Reforms Needed to Protect Skilled American Workers." Mar. 17. Accessed Apr. 8, 2015. http://www.judiciary.senate.gov/imo/media/doc /Hira%20Testimony.pdf.

Hira, Ron, and Anil Hira. 2008. *Outsourcing America: The True Cost of Shipping Jobs Overseas and What Can Be Done about It*. New York: AMACOM.

Howard, Lee. 2008. "Pfizer to Ax IT Contractors?" *The Day*, Nov. 3.

IBM. 2008. Form 10-K Annual Report. Dec. 31. Accessed Apr. 8, 2015. https:// www.sec.gov/Archives/edgar/data/51143/000104746909001737/0001047469 -09-001737-index.htm.

Infosys. 2005. "Infosys Technologies Limited Earnings Conference Call (US)." July 12. Accessed Apr. 8, 2015. http://www.infosys.com/investors/reports-filings /quarterly-results/2005-2006/Q1/Documents/transcripts/USEarningsconference -12-07-05.pdf.

———. 2009. Q210 Results, Presentation Made at Press Conference, Results for the Second Quarter ended Sept. 30, 2009. Accessed Apr. 8, 2015. http://www.infosys .com/investors/reports-filings/quarterly-results/2009-2010/Q2/Documents/press -conference-Q2-10.pdf.

Jasso, Guillermina, Vivek Wadhwa, Gary Gereffi, Ben Rissing, and Richard Freeman. 2010. "How Many Highly Skilled Foreign-Born are Waiting in Line for U.S. Legal Permanent Residence?" *International Migration Review* 44:477–98.

Kruse, Michael, and Theresa Blackwell. 2008. "How Oldsmar Got Global Influence." *Tampa Bay Times*, Sept. 19. http://www.tampabay.com/news/business/how -oldsmar-got-global-influence/818379.

Lohr, Steve. 2007. "Parsing the Truths about Visas for Tech Workers." *New York Times*, Apr. 15.

———. 2009. "Piecemeal Layoffs Avoid Warning Laws." *New York Times*, Mar. 5.

Lowell, B. Lindsay. 2000. "H-1B Temporary Workers: Estimating the Population." Working Paper no. 12, Center for Comparative Immigration Studies, University of California, San Diego.

Marshall, Ray. 2009. *Immigration for Shared Prosperity: A Framework for Comprehensive Reform*. Washington, D.C.: Economic Policy Institute.

McDougall, Paul. 2006. "Analyst: IBM to Employ 100,000 Workers in India by 2010." *InformationWeek*, Dec. 4.

McGee, Marianne Kolbasuk. 2007. "With the H-1B Visa Cap Filled in Record Time,

Reform is in the Air." *InformationWeek*, Apr. 7. Accessed Apr. 8, 2015. http://
www.informationweek.com/news/global-cio/showArticle.jhtml?articleID=1988
00918&pgno=3.

NPR Marketplace Radio. 2007. "H-1B Visa Just a Ticket to the Way Station."
National Public Radio, July 30. Accessed Apr. 8, 2015. http://marketplace.publi-
cradio.org/display/web/2007/07/30/h1b_visa_just_a_ticket_to_the_way_station/.

Singh, Shelley. 2003. "U.S. Visas are not a TCS-Specific Issue." *Businessworld*, June.

Srivastava, Mehul, and Moira Herbst. 2010. "The Return of the Outsourced Job."
Businessweek, Jan. 11.

Toppo, Gregg, and Icess Fernandez. 2009. "Federal Complaint: Filipino Teachers
Held in 'Servitude.'" *USA Today*, Oct. 27. Accessed Nov. 4, 2009. http://www
.usatoday.com/news/education/2009-10-27-filipino-teachers_N.htm.

U.S. Department of Homeland Security. 2012. "Yearbook of Immigration Statistics:
2008." Table 7. Persons Obtaining Legal Permanent Resident Status by Type and
Detailed Class of Admission: Fiscal Year 2012. Accessed Apr. 8, 2015. http://
www.dhs.gov/sites/default/files/publications/immigration-statistics/yearbook
/2012/LPR/table7d.xls.

———. 2015. "Characteristics of H-1B Specialty Occupation Workers. Fiscal Year
2014 Annual Report to Congress." Feb. 26. https://www.uscis.gov/sites/default
/files/USCIS/Resources/Reports%20and%20Studies/H-1B/h-1B-characteristics
-report-14.pdf.

U.S. State Department. 2015. "Visa Bulletin for February 2015." Accessed Apr. 8,
2015. http://travel.state.gov/content/visas/english/law-and-policy/bulletin/2015
/visa-bulletin-for-february-2015.html.

Washington Post Editorial. 2008. "A Recipe for Weakness." *Washington Post*,
June 4. http://www.washingtonpost.com/wp-dyn/content/article/2008/06/03
/AR2008060303102.html.

Contributors

anthony lising antonio
School of Education
Stanford University
Stanford, CA 94305

Stephen R. Barley
Technology Management Program
College of Engineering
University of California, Santa
    Barbara
1320 Phelps Hall
Santa Barbara, CA 93106-5129

Erling Barth
Institute for Social Research
PO Box 3233 Elisenberg
N-0208 Oslo, Norway

Samantha R. Brunhaver
The Polytechnic School
Ira A. Fulton Schools of Engineering
Arizona State University
7171 E. Sonoran Arroyo Mall
Mesa, AZ 85212

Helen L. Chen
Designing Education Lab
Office of the Registrar, Wallenberg Hall
Stanford University
450 Serra Mall, Building 160
Stanford, CA 94305-2055

James C. Davis
Boston Federal Statistical Research
    Data Center
National Bureau of Economic
    Research
1050 Massachusetts Avenue
Cambridge, MA 02138

Richard B. Freeman
National Bureau of Economic
    Research
1050 Massachusetts Avenue
Cambridge, MA 02138

Shannon K. Gilmartin
The Michelle R. Clayman Institute for
    Gender Research
Stanford University
589 Capistrano Way
Stanford, CA 94305-8640

Susan Helper
Weatherhead School of Management
Case Western Reserve University
11119 Bellflower Road
Cleveland, OH 44106-7235

Ron Hira
Department of Political Science
Howard University
Washington, DC 20059

Yoon Sun Hur
Korea Institute for International
    Economic Policy (KIEP)
[30147] Building C, Sejong National
    Research Complex, 370
Sicheong-Daero, Sejong-si, Korea

Morris M. Kleiner
University of Minnesota
Humphrey School of Public Affairs
260 Humphrey Center
301 19th Street South
Minneapolis, MN 55455

Russell F. Korte
Department of Human Organizational
    Learning
Graduate School of Education and
    Human Development
The George Washington University
2134 G St NW
Washington, DC 20037

Jennifer Kuan
A. B. Freeman School of Business
Goldring/Woldenberg Hall
Tulane University
7 McAlister Drive
New Orleans, LA 70118-5698

Daniel Kuehn
The Urban Institute
2100 M Street, NW
Washington, DC 20037

Leonard Lynn
Weatherhead School of Management
Case Western Reserve University
10900 Euclid Avenue
Cleveland, OH 44106-7235

Hal Salzman
E. J. Bloustein School of Planning &
    Public Policy
J. J. Heldrich Center for Workforce
    Development
Rutgers University
New Brunswick, NJ 08901

Sheri D. Sheppard
Department of Mechanical
    Engineering
Peterson Building (550)
Stanford University
Stanford, CA 94305-4021

Andrew J. Wang
National Bureau of Economic
    Research
1050 Massachusetts Avenue
Cambridge, MA 02138

Yingchun Wang
Davies College of Business
University of Houston, Downtown
One Main Street
Houston, TX 77002

Catherine J. Weinberger
Institute for Social, Behavioral and
    Economic Research (ISBER)
University of California
Santa Barbara, CA 93106-2150

Author Index

Subject Index

Page numbers followed by "f" or "t" refer to figures or tables respectively.